Definitions for Hardware and Software Safety Engineers

Springer-Verlag London Ltd.

Meine van der Meulen

Definitions for Hardware and Software Safety Engineers

Springer

Meine van der Meulen
Simtech Engineering BV, Max Euwelaan 60, 3062 MA, Rotterdam, The Netherlands

ISBN 978-1-4471-1148-1

British Library Cataloguing in Publication Data
Meulen, Meine van der
Definitions for hardware and software safety engineers
1.System safety - Terminology 2.Computer security - Terminology 3.Computer engineering - Terminology
I.Title
621.3'9'0289
ISBN 978-1-4471-1148-1 ISBN 978-1-4471-0471-1 (eBook)
DOI 10.1007/978-1-4471-0471-1

Library of Congress Cataloging-in-Publication Data
A catalog record for this book is available from the Library of Congress

Originally published by Springer-Verlag London Limited in 2000
Softcover reprint of the hardcover 1st edition 2000

Typesetting: by Florence Production Ltd.

34/3830-543210 Printed on acid-free paper SPIN 10724630

Preface

Every science has its own vocabulary. This is a prerequisite for communication. In a relatively new science like safety engineering of electrotechnical systems, the vocabulary develops. As insight into the meaning of the phenomena grows, the terms that describe them get better defined. However, without sufficient communication between scientists, the definitions might drift apart. Similar terms may then describe different phenomena, and similar phenomena may be described by different terms.

This dictionary is a compilation of definitions as they appear in literature. The citation is literal, except for standardisation of the enumeration style, minor alterations, and conversion to Oxford English spelling. Also, most examples and notes in definitions have been deleted.

For obvious reasons it is not possible to quote all definitions from all sources. I have used the following selection criteria to include a definition from a source in this dictionary.

First, a term has to fit within the scope of the dictionary: electrotechnical systems safety. This means all terms directly associated to the safe use and analysis of electric, electronic and programmable electronic systems. The dictionary, however, contains almost no terms on statistics, security and quality assurance, although some might argue that they fit within its scope. Further, the reader might notice a slight bias towards the engineering side of electrotechnical systems safety, which is my own field of interest.

Second, there appears to be some sort of hierarchy in sources. International standards are of most importance; scientific publications are mostly of less importance. In between there is some sort of order that is hard to make explicit. In some cases considerations like this influence the selection of a definition. If, for example, two or more sources define a term slightly differently, I cite the one from the most important source. On the other hand I prefer to have at least two definitions of a term, thus giving way to less important definitions where a second definition would otherwise not be available.

The order of listing of the definitions of a term is by year of publication.

The fact that the dictionary literally cites from English sources only biases its contents to the views on safety engineering in English-speaking countries. Although I do not think that this may lead to significant problems, the user should be aware of this.

I hope this dictionary will help its users to use the right terms in the right way.

Meine van der Meulen
Rotterdam, September 1999

Acknowledgements

Material has been cited with the permission of DIN Deutsches Institut für Normung e.V. and VDE Verband der Elektrotechnik Elektronik Informationstechnik e.V. Normative for the use of standards are the versions with the latest date of publication. Copies of these standards can be obtained from VDE-Verlag GmbH, Bismarckstr. 33, 10625 Berlin und der Beuth Verlag GmbH, Burggrafenstr. 6, 10787 Berlin.

Various definitions as indicated are excerpted from various IEEE Standards which are detailed in the References at the end of this book. All such material is copyrighted by The Institute of Electrical and Electronics Engineers, Inc. The IEEE disclaims any responsibility or liability resulting from the placement and use in this publication. Information is reprinted with the permission of the IEEE.

Further definitions as indicated are excerpted from various IEC publications which are detailed in the References at the end of this book. All such material is copyrighted to the IEC and reprinted with permission. The IEC disclaims any responsibility or liability resulting from the placement and use of this information in this book. Copies of the IEC publications can be obtained from: Customer Service Centre, IEC Central Office, 3 rue de Varembé, 1211 Geneva 20, Switzerland. Tel +41 22 919 02 11, Fax +41 22 919 03 00, Email custserv@iec.ch, Website http://www.ies.ch

The author would like to thank all those who kindly gave their permission to use material within this book. While the author has made considerable efforts to ensure that each definition is cited correctly, he cannot accept any legal responsibility or liability for any errors or omissions.

A

Abatement, Risk

See Risk Abatement.

Abend

Abbreviation for Abnormal End (IEEE Std–610.12, 1991).

Abnormal End

Abend. Termination of a process prior to completion (IEEE Std–610.12, 1991).

Abort

To terminate a process prior to completion (IEEE Std–610.12, 1991).

Abstraction

A view of an object that focuses on the information relevant to a particular purpose and ignores the remainder of the information (UK MoD, Def Stan 00–55, 1997).

Acceptance

An action by an authorised representative of the acquirer by which the acquirer assumes ownership of software products as partial or complete performance of a contract (USA DoD, MIL-Std–498, 1994).

A conclusion that a batch, lot or quantity of product or service satisfies the acceptance criteria (BSI, BS 4778–2, 1991).

Acceptance, Risk

See Risk Acceptance.

Acceptance Test

A test conducted under specified conditions by, or on behalf of, the government, using delivered or deliverable items, in order to determine the item's compliance with specified requirements (includes acceptance of first production units) (USA DoD, MIL-Std-721C, 1992).

A contractual test to prove to the customer that the device meets certain conditions of its specification, as stated in the contract (IEC 902, 1987).

Acceptance Test, Pre-Startup

See Pre-Startup Acceptance Test.

Acceptance Testing

Formal testing conducted to enable a user, customer, or other authorised entity to determine whether to accept a system or component (IEEE Std-610.12, 1991).

Formal testing conducted to determine whether or not a system satisfies its acceptance criteria and to enable the customer to determine whether or not to accept the system (IEEE Std-610.12, 1991; IEEE Std-1012, 1987; IEEE Std-729, 1983).

Testing carried out specifically to demonstrate to the user that the requirements of the functional specification have been met (Smith & Wood, 1989).

Accessibility

A measure of the relative ease of admission to the various areas of an item for the purpose of operation or maintenance (NATO, ARMP-7, 1996; USA DoD, MIL-Std-721C, 1992).

Accident

An unintended event or series of events that result in death, injury, loss of a system or service, or environmental damage (CENELEC, ENV50129, 1998).

An undesired event arising from operation of any project-specific items which results in: (1) human death or injury; (2) loss of, or damage to, project hardware, software or facilities which could then affect the accomplishment of the mission; (3) loss of, or damage to, public or private property; or (4) detrimental effects on the environment (ESA, ECSS-P-001A, 1997).

An unintended event or sequence of events that causes death, injury, environmental or material damage (UK MoD, Def Stan 00-56, 1996).

An accident is an undesired and unplanned (but not necessarily unexpected) event that results in (at least) a specified level of loss (Leveson, 1995).

An unplanned event or series of events that result in death, injury, illness, environmental damage, or damage to or loss of equipment or property (IEEE Std-1228, 1994).

Accident, Near

See Near Accident.

Accident Sequence

The progression of events that result in an accident. This is illustrated in the following diagram: initiating event → hazardous state → accident (UK MoD, Def Stan 00-56, 1996).

Accidental Fault

Fault appearing or created fortuitously (Laprie, 1992).

Accreditation

(1) The procedure for accepting an IT system for use within a particular environment. (2) The procedure for recognising both the technical competence and the impartiality of a test laboratory to carry out its associated tasks (UK DTI, ITSEC, 1991).

Accreditation, Security

See Security Accreditation.

Accuracy

(1) A qualitative assessment of correctness, or freedom from error. (2) A quantitative measure of the magnitude of error (IEEE Std–610.12, 1991).

Achieved

Obtained as the result of measurement (USA DoD, MIL-Std–721C, 1992).

Act of God

An accident causal factor generally interpreted as being beyond human control (ASSE, 1988).

Action, Corrective

See Corrective Action.

Action, Maintenance

See Maintenance Action.

Active Failure

Failure of an item which causes the operation of the protection devices around it and results in the opening of one or more circuit breakers or blowing of one or more fuses (IEC 50–191, Am. 1, 1999).

Active Fault

Fault which causes control actions although the conditions specified in the program are not fulfilled (IEC 50–351, 1998).

Fault producing an error (Laprie, 1992).

Active Inventory

The group of items assigned to an operational status (USA DoD, MIL-Std–721C, 1992).

Active Redundancy

That redundancy wherein all redundant items are operating simultaneously (O'Connor adds: rather than being switched on when needed) (USA DoD, MIL-Std–721C, 1992; O'Connor, 1981).

In fault tolerance, the use of redundant elements operating simultaneously to prevent, or permit recovery from, failures (IEEE Std–610.12, 1991).

Active Repair Time

The average time required to analyse the failure, repair and return the item to a state of readiness (OREDA, 1992).

The portion of down-time during which repairmen are actively repairing the system, which includes set-up time, fault diagnosis time, fault correction time, and final checkout time (Martz & Waller, 1982).

Active Time

That time during which an item is in an operational inventory (USA DoD, MIL-Std–721C, 1992).

Actuator

A device that converts a signal into a corresponding movement (IEC 50–351, 1998).

Adaptive Maintenance

Modification of a software product performed after delivery to keep a computer program usable in a changed or changing environment (IEEE Std–1219, 1992).

Software maintenance performed to make a computer program usable in a changed environment (IEEE Std–610.12, 1991).

Adaptive Software Enhancement

Software changes necessary to accommodate change(s) in the operational or hardware environment (USA DoD, MIL-Hdbk–347, 1990).

Adjust

To restore to a specification without rework or replacement of any entity (UK MoD, Def Stan 00–49, 1996).

Adjustment

Changing (by electronic, electrical or physical means) a variable in an item to cause a change in its output characteristics (NATO, ARMP–7, 1996).

Administrative Time

That element of delay time, not included in the supply delay time (USA DoD, MIL-Std–721C, 1992).

The portion of down-time not included under active repair time and logistic time, which includes the time spent preparing orders for repair, waiting for repairmen, etc. (Martz & Waller, 1982).

ALARA

As Low As Reasonably Achievable (ASSE, 1988).

Alarm Unit

A device with an audible and/or visible output which indicates an abnormal or out-of-limit condition in a plant or control system (IEC 902, 1987).

ALARP

As Low As Reasonably Practicable.

Alert

The formal notification to users, informing them of failures or non-conformances of items, already released for use or not, which could also be present on other items already delivered (e.g. items with identical design concept, materials, components or processes) (ESA, ECSS-P-001A, 1997).

Alert Time

That element of up-time during which an item is assumed to be in specified operating condition and is awaiting a command to perform its intended mission (NATO, ARMP-7, 1996; USA DoD, MIL-Std-721C, 1992).

Algorithm

A completely determined finite sequence of instructions by which the values of the output variables can be calculated from the values of the input variables (IEC 50-351, 1998).

A finite ordered set of well-defined rules for the solution of a problem (ISO 2382-1, 1993).

A finite set of well-defined rules that gives a sequence of operations for performing a specific task (RTCA, DO178B, 1992).

(1) A finite set of well-defined rules for the solution of a problem in a finite number of steps. (2) Any sequence of operations for performing a specific task (IEEE Std-610.12, 1991).

A set of logical rules for the solution of a defined problem (Smith & Wood, 1989).

A set of well-defined rules for the solution of a problem in a finite number of steps (IEC 902, 1987).

Alignment

Performing the adjustments that are necessary to return an item to specified conditions (USA DoD, MIL-Std-721C, 1992).

Allocated Baseline

The initially approved documentation describing the item's functional and interface characteristics that are allocated from those of a higher-level CI, interface requirements with interfacing configuration items, additional design constraints and the verification required to demonstrate the achievement of those specified functional and interface characteristics (USA DoD, MIL-Hdbk-347, 1990; USA DoD, MIL-Std-480B, 1988).

Allocation

A form of co-ordination control that assigns a resource to a batch or unit (ISA, S88.01, 1995).

A procedure applied during the design of an item intended to apportion the requirements for performance measures for an item to its subitems according to given criteria (IEC 300-3-1, 1991).

Alpha Test

The first functional test of a device or system (CCPS, 1993).

ALT

Accelerated Life Test.

Analysis

The determination of the essential qualities, performance and limitations of an item by cognitive or computational methods (ESA, ECSS-P-001A, 1997).

Examination for the purpose of understanding (IEEE Std-1074, 1991).

Analysis, Cause-Consequence

See Cause-Consequence Analysis.

Analysis, Event Tree

See Event Tree Analysis.

Analysis, Failure

See Failure Analysis.

Analysis, Failure Mode and Effects

See Failure Mode and Effects Analysis.

Analysis, Fault Tree

See Fault Tree Analysis.

Analysis, Impact

See Impact Analysis.

Analysis, Parts Count

See Parts Count Analysis.

Analysis, Product

See Product Analysis.

Analysis, Safety

Safety Analysis.

Analysis, Sneak Circuit

See Sneak Circuit Analysis.

Analysis, Static

See Static Analysis.

Animation

Simulated operation of the software system (or of some significant portion of the system) to display significant aspects of the behaviour of the system, for instance applied to a requirements specification in an appropriate format or an appropriate high-level representation of the system design (IEC 61508-4, 1998).

Annunciator

An indicating device that displays a legend illuminated internally on receipt of an initiating signal (IEC 902, 1987).

Anomaly

Any deviation from the expected situation (ESA, ECSS-P-001A, 1997).

Any deviation from requirements, expected or desired behaviour, or performance of the software (IEEE Std-1074, 1991).

Anything observed in the documentation or operation of software that deviates from expectations based on previously verified software products or reference documents (IEEE Std-610.12, 1991; IEEE Std-1012, 1987).

Anomaly, Critical

See Critical Anomaly.

Application Hardware

A special-purpose unit designed, as a peripheral to the computer, to carry out some specified function (Smith & Wood, 1989).

Application Language

A problem-oriented language whose statements closely resemble the jargon or terminology of the particular problem type (Smith & Wood, 1989).

Application Program

See Software (ISA, S84.01, 1996).

A logical assembly of all the programming language elements and constructs necessary for the intended signal processing required for the control of a machine or process by a Programmable Controller System (IEC 1131-1, 1992).

Application Software

All software used to operate, or that is resident in, a target computer system (USA DoD, MIL-Hdbk-1467, 1997).

Software specific to the user application in that it is the SIS functional description programmed in the PES to meet the overall Safety Requirement Specifications. In general, it contains logic sequences, permissives, limits, expressions, etc., that control the appropriate input, output, calculations, decisions necessary to meet the safety functional requirements (ISA, S84.01, 1996).

Software or a program that is specific to the solution of an application problem (ISO 2382-1, 1993).

Software designed to fulfil specific needs of a user (IEEE Std-610.12, 1991).

The software written for a computer for the purpose of applying the computer functions to solve a particular problem. This is distinct from the resident operating software which is part of the computer system (Smith & Wood, 1989).

Apportionment

A process whereby the RAMS elements for a system are subdivided between the various items which comprise the system to provide individual targets (CENELEC, prEN50126, 1998).

The determination and assignment of the reliability requirements to any entity (UK MoD, Def Stan 00-49, 1996).

Appraisal Cost

Costs incurred while conducting inspections, tests, and other planned evaluations to insure requirements are met (USA DoD, MIL-Std-109C, 1994).

The cost of assessing the quality achieved (BSI, BS 4778-2, 1991).

Approval

Formal agreement to use or apply an item (ESA, ECSS-P-001A, 1997).

Permission for a product, process or service to be marketed or used for stated purposes or under stated conditions (ISO/IEC, Guide 2, 1996).

Written notification by an authorised representative of the acquirer that a developer's plans, design, or other aspects of the project appear to be sound and can be used as the basis for further work. Such approval does not shift responsibility from the developer to meet contractual requirements (USA DoD, MIL-Std-498, 1994).

The act or instance of expressing a favourable opinion or giving formal or official sanction (RTCA, DO178B, 1992).

An act of endorsing or adding positive authorisation or both (IEC 902, 1987).

Approval, Design

See Test Qualification (USA DoD, MIL-Std-721C, 1992).

Approval, Type

See Type Approval.

Approved, User

See User Approved.

Arbitrary failure

See Failure (Laprie, 1992).

Arbitration

A form of co-ordination control that determines how a resource should be allocated when there are more requests for the resource than can be accommodated at one time (ISA, S88.01, 1995).

Architecture

Specific configuration of hardware and software elements in a system (IEC 61508-4, 1998).

The arrangement and interconnection of the hardware components or modules that comprise the Safety Instrumented System (ISA, S84.01, 1996).

The structure and connectivity of the elements composing a complex system (SSCP 85, 1995).

The organisational structure of a system or component (IEEE Std-610.12, 1991).

Architecture, Computer

See Computer Architecture.

Argument, Formal

See Formal Argument.

Argument, Rigorous

See Rigorous Argument.

Assemble

To translate from an assembly language program into an object language (ISO/IEC DIS 2382-7, 1996).

Assembler

A program for converting instructions, written in mnemonics, into binary machine code suitable to operate a computer or other programmable device (Smith & Wood, 1989).

A computer program that converts or translates a symbolic language program into a machine language form so that the computer can handle or process problem data (ASSE, 1988).

Assembly Language

A machine-oriented language that provides symbolic naming of operations and locations and other features such as macro-instructions (ISO/IEC DIS 2382-7, 1996).

A low-level language where the instructions are expressed as mnemonics and where each instruction corresponds to a simple computer function (Smith & Wood, 1989).

Assessment

The process of analysis to determine whether the Design Authority and the validator have achieved a product that meets the specified requirements and to form a judgement as to whether the product is fit for its intended purpose (CENELEC, ENV50129, 1998).

The undertaking of an investigation in order to arrive at a judgement, based on evidence, of the suitability of a product (CENELEC, prEN50126, 1998).

The result of a systematic examination and judgement of something against stated criteria or requirements (BSI, BS 4778-3.1, 1991).

Assessment, Functional Safety

See Functional Safety Assessment.

Assessment, Safety

See Safety Assessment.

Assessor

The person or agent appointed to carry out the assessment (CENELEC, ENV50129, 1998).

Assurance

The planned and systematic actions necessary to provide adequate confidence and evidence that a product or process satisfies given requirements (RTCA, DO178B, 1992).

The confidence that may be held in the security provided by a Target of Evaluation (UK DTI, ITSEC, 1991).

Assurance, Quality

See Quality Assurance.

Assurance, Safety

See Safety Assurance.

Atomic System

System whose internal structure cannot be discerned, or is not of interest and can be ignored (Laprie, 1992).

ATP

Acceptance Test Procedure.

Attribute

A property of an entity, either physical or logical (UK MoD, Def Stan 00-58, 1996).

Audit

A systematic and independent examination to determine whether the procedures specific to the requirements of a product comply with the planned arrangements, are implemented effectively and are suitable to achieve the specified objectives (CENELEC, prEN50126, 1998).

An independent examination of the life cycle processes and their products for compliance, accuracy, completeness, consistency and traceability (RTCA, DO178B, 1992).

A defined searching examination (BSI, BS 4778–3.1, 1991).

An independent examination of a work product or set of work products to assess compliance with specifications, standards, contractual agreements, or other criteria (IEEE Std–610.12, 1991).

An independent evaluation of software products or processes to ascertain compliance to standards, guidelines, specifications, and procedures based on objective criteria that include documents that specify: (1) The form or content of the products to be produced. (2) The process by which the products shall be produced. (3) How compliance to standards or guidelines shall be measured (IEEE Std–1028, 1988).

An activity to determine, through investigation, the adequacy of an adherence to established procedures, instructions, specifications, codes and standards or other applicable contractual and licensing requirements, and the effectiveness of implementation (IEC 902, 1987).

Audit, Configuration

See Configuration Audit.

Audit, Functional Safety

See Functional Safety Audit.

Audit, Process Safety

See Process Safety Audit.

Audit, Safety

See Safety Audit.

Authorisation

The formal permission to use a product within specified application constraints (CENELEC, ENV50129, 1998).

Automated System

A control system beyond the scope of IEC 1131 in which Programmable Controller Systems are incorporated by or for the user, but which also contains other components including their application programs (IEC 1131-1, 1992).

Automated Verification System

(1) A software tool that accepts as input a computer program and a representation of its specification and produces, possibly with human help, a proof or disproof of the correctness of a program. (2) Any software tool that automates part or all of the verification process (IEEE Std-610.12, 1991).

Automatic Test

Equipment for stimulus and measurement controlled by a programmed sequence of steps (usually in software) (Smith, 1981).

Autotest

See Built-In-Test (NATO, ARMP-7, 1996).

Availability

The ability of a product to be in a state to perform a required function under given conditions at a given instant of time or over a given time interval assuming that the required external resources are provided (CENELEC, ENV50129, 1998; CENELEC, prEN50126, 1998).

The ability of a functional unit to be in a state to perform a required function under given conditions at a given instant of time or over a given time interval, assuming that the required external resources are provided (ISO 2382-14, 1997).

The ability of an item to be in a state to perform a required function under given conditions at a given instant of time or over a given time interval, assuming that the required external resources are provided (ESA, ECSS-P-001A, 1997; BSI, BS 4778-3.2, 1991; IEC 50-191, 1990).

The probability that the system or equipment used under stated conditions will be in an operable and committable state at any given time (NATO, ARMP-7, 1996).

See Safety Availability (ISA, S84.01, 1996).

A measure of the degree to which an item is in an operable and committable state at the start of a mission when the mission is called for at an unknown (random) time. The ability of an item to perform its designated function when required for use (USA DoD, MIL-Std-109C, 1994).

The probability that a system will be able to perform its designated function when required for use (CCPS, 1993).

Dependability with respect to the readiness for usage. Measure of correct service delivery with respect to the alternation of correct and incorrect service (Laprie, 1992).

The fraction of time that the system is actually capable of performing its mission (IEC 1131-1, 1992).

The degree to which a system or component is operational and accessible when required for use. Often expressed as a probability (IEEE Std-610.12, 1991).

The prevention of the unauthorised withholding of information or resources (UK DTI, ITSEC, 1991).

Expected fraction of time during which a system or component is functioning acceptably (Musa et al., 1987).

The probability that the system is operating satisfactorily at any point in time when operating under specified conditions, where the time categories considered include all but free time (Martz & Waller, 1982).

The ability of an item (under combined aspects of its reliability, maintainability and maintenance support) to perform its required function at a stated instant of time or over a stated period of time (O'Connor, 1981).

Availability, Effective

See Effective Availability.

Availability, Intrinsic

See Intrinsic Availability.

Availability, Operational

See Operational Availability.

Availability, Safety

See Safety Availability.

Availability, Steady

See Steady Availability.

Avoidance, Fault

See Fault Avoidance.

B

Back-to-Back Testing

Testing in which two or more variants of a program are executed with the same inputs, the outputs are compared, and errors are analysed in case of discrepancies (IEEE Std–610.12, 1991).

Backup

(1) A system, component, file, procedure, or person available to replace or help restore a primary item in the event of a failure or externally caused disaster. (2) To create or designate a system, component, file, procedure, or person as in (1) (IEEE Std-610.12, 1991).

Backward Recovery

Form of error recovery where the erroneous state transformation consists of bringing the system back to previously occupied state (Laprie, 1992).

(1) The reconstruction of a file to a given state by reversing all changes made to the file since it was in that state. (2) A type of recovery in which a system, program, database, or other system resource is restored to a previous state in which it can perform required functions (IEEE Std-610.12, 1991).

Baseline

Set of information which describes exhaustively a situation at a given instant of time or over a given time interval (ESA, ECSS-P-001A, 1997).

The agreed specification, or software item, which has been uniquely identified and becomes the focus for further development, and which can only be altered under strict control procedures (IEEE Std-1298, 1992; SA, AS 3563, 1991).

The approved, documented configuration of a software or hardware configuration item, that thereafter serves as the basis for further development, and can be changed only through change control procedures (RTCA, DO178B, 1992).

(1) A work product that has been formally reviewed and agreed upon, and that can be changed only through formal change control procedures (only this definition in IEEE Std-1058.1). (2) A document or a set of such documents formally designated and fixed at a specific time during the life cycle of a configuration item. (3) Any agreement or result designated and fixed at a given time, from which changes require justification and approval (IEEE Std-610.12, 1991; IEEE Std-1058.1, 1988).

A configuration identification document or a set of such documents formally designated by the government at a specific time during a CI's life cycle (USA DoD, MIL-Hdbk-347, 1990; USA DoD, MIL-Std-480B, 1988).

Baseline, Allocated

See Allocated Baseline.

Baseline, Configuration

See Configuration Baseline.

Baseline, Functional

See Functional Baseline.

Baseline, Product

See Product Baseline.

Baseline, Software

See Software Baseline.

Basic Control

Control that is dedicated to establishing and maintaining a specific state of equipment or process condition (ISA, S88.01, 1995).

Basic Process Control System

A system that responds to input signals from the equipment under control and/or from an operator and generates output signals, causing the equipment under control to operate in the desired manner (ISA, S84.01, 1996).

The control equipment and system which is installed to regulate normal production functions (CCPS, 1993).

Batch

(1) The material that is being produced or that has been produced by a single execution of a batch process. (2) An entity that represents the production of a material at any point in the process (ISA, S88.01, 1995).

Pertaining to a system or mode of operation in which inputs are collected and processed all at one time, rather than being processed as they arrive, and a job, once started, proceeds to completion without additional input or user interaction (IEEE Std–610.12, 1991).

Batch Control

Control activities and control functions that provide a means to process finite quantities of input materials by subjecting them to an ordered set of processing activities over a finite period of time using one or more pieces of equipment (ISA, S88.01, 1995).

Batch Process

A process that leads to the production of finite quantities of material by subjecting quantities of input materials to an ordered set of processing activities over a finite period of time using one or more pieces of equipment (ISA, S88.01, 1995).

Batch Processing

The technique of executing a set of computer programs so that each program is completed before the next one of the set is started (ASSE, 1988).

Bathtub Curve

A graph of the number of failures in a system or component as a function of time. The name is derived from the usual shape of the graph: a period of decreasing failures (the early-failure period), followed by a relatively steady period (the constant-failure period), followed by a period of increasing failures (the wearout-failures period) (IEEE Std–610.12, 1991).

Behaviour

What a system does (Laprie, 1992).

Benchmark

(1) A standard against which measurements or comparisons can be made. (2) A procedure, problem, or test that can be used to compare systems or components to each other or to a standard as in (1). (3) A recovery file (IEEE Std–610.12, 1991).

Benign Failure

Failure whose penalties are of the same order of magnitude as the benefit provided by correct service delivery (Laprie, 1992).

Beta Test

A device or system test comprising production components that have satisfactorily passed an alpha test. The scale of the test is such that the equipment is installed in an operating environment and its operation is monitored to determine performance against expectations (CCPS, 1993).

Big-Bang Testing

A type of integration testing in which software elements, hardware elements, or both are combined all at once into an overall system, rather than in stages (IEEE Std–610.12, 1991).

Binding

A set of interface specifications which allows access to a third-party item of software, possibly a library, usually written in another programming language (SSP 84, 1996).

Representations of interfaces in terms of standard programming languages (SSCP 85, 1995).

Bit Error Rate

The rate of incidence of random incorrect binary bits. This usually refers to the effect of corruption in a communication channel (Smith & Wood, 1989).

The random incidence of incorrect binary digits (Smith, 1981).

BITE

See Built-In Test Equipment.

Black Box

(1) A system or component whose inputs, outputs, and general functions are known but whose contents or implementation are unknown or irrelevant. (2) Pertaining to an approach that treats a system or component as in (1) (IEEE Std–610.12, 1991).

A generic term used to describe a device which performs a certain function for which only the input and output signals are known (IEC 902, 1987).

Black Box Test

See Functional Testing (IEEE Std–610.12, 1991).

Black Box Testing

Techniques for test case generation which do not rely on knowledge of the internal structures of the item under test (UK MoD, Def Stan 00–55, 1997).

Block Diagram, Reliability

See Reliability Block Diagram.

Block Structured Language

A design or programming language in which sequences of statements, called blocks, are defined, usually with 'begin' and 'end' delimiters, and variables or labels defined in one block are not recognised outside that block (UK MoD, Def Stan 00-55, 1997).

Boundary

The interface between an item and its surroundings (OREDA, 1992).

Box, Black

See Black Box.

Box, Glass

See Glass Box.

Box, White

See White Box.

BPCS

See Basic Process Control System.

Branch Testing

Testing designed to execute each outcome of each decision point in a computer program (IEEE Std-610.12, 1991).

Breakdown Criterium

Limiting levels of the permissible value for the deviation form its characteristic properties of a component after demands start to be made on it (Hölscher & Rader, 1986).

Breakpoint

A point in a program, module, or statement where execution may be suspended depending on a specific condition or event (ISO/IEC DIS 2382-7, 1996).

A point in a computer program at which execution can be suspended to permit manual or automated monitoring of program performance or results (IEEE Std-610.12, 1991).

Brownout

The result of the deliberate lowering of voltage (and, thus, the power supplied to all users) by an electric utility; employed when demand for power exceeds generating capacity (ASSE, 1988).

Bug

See Error, Fault (IEEE Std-610.12, 1991).

A slang expression for a software fault (Smith & Wood, 1989).

Bug Seeding

See Error Seeding (ISO/IEC DIS 2382-7, 1996; IEEE Std-610.12, 1991).

Built-In-Test

An integral capability of the equipment which provides an on-board test capability to detect, diagnose, or isolate system failures. The fault detection and, possibly, isolation capability is used for periodic or continuous monitoring of a system's operational health, and for observation and, possibly, diagnosis as a prelude to maintenance action (NATO, ARMP-7, 1996).

Built-In Test Equipment

BITE. Any device permanently mounted in the equipment and used for the express purpose of testing the equipment, either independently or in association with external test equipment (NATO, ARMP-7, 1996).

Burn-In

(1) A process of increasing the reliability of a new or refurbished repairable functional unit by operating it in a prescribed environment, detecting as many early faults as possible, and eliminating them by

corrective maintenance. (2) A screening test employing the functional operation of a non-reparable functional test (ISO 2382-14, 1997).

The operation of an item under stress to stabilise its characteristics (NATO, ARMP-7, 1996; USA DoD, MIL-Std-721C, 1992).

A process of increasing the reliability performance of hardware employing functional operation of every item in a prescribed environment with successive corrective maintenance at every failure during the early failure period (BSI, BS 4778-3.2, 1991; IEC 50-191, 1990).

The operation of items for a specified period of time in order to remove early failures and bring the reliability characteristic into the random failure part of the bathtub curve (Smith, 1981).

Burn-In Period

See Early Failure Period (IEEE Std-610.12, 1991).

Busy

Pertaining to a system or component that is operational, in service, and in use (IEEE Std-610.12, 1991).

Bypassing

Act of temporarily defeating a safety function in a Safety Instrumented System (ISA, S84.01, 1996).

C

Calendar Time

The period of time between the dates of data surveillance for one particular item (OREDA, 1992).

Chronological time, including time during which a computer may not be running (Musa et al., 1987).

Calibration

All the operations for the purpose of determining the values of the errors and, if necessary, other metrological properties of a measuring instrument (ESA, ECSS-P-001A, 1997; IEC 50-301, 1983).

A comparison of a measuring device with a known standard. Not to be confused with alignment (USA DoD, MIL-Std-721C, 1992).

Cannibalise

Robbing. To remove, without subsequent replacement, useful serviceable items from one equipment/system for use on other equipment/systems (UK MoD, Def Stan 00-49, 1996).

Cascade Tripping

Sequential forced tripping of generation and/or transmission units caused by excessive loading of lines carrying power into or out of an area (IEC 50-191, Am. 1, 1999).

CASE

Computer Aided Software Engineering.

Case Generator, Test

See Test Case Generator.

Case Specification, Test

See Test Case Specification.

Case, Test

See Test Case.

Catastrophe

A loss of extraordinarily large dimensions in terms of injury, death, damage, and destruction (ASSE, 1988).

Catastrophic Failure

Failure whose consequences are incommensurably greater than the benefit provided by correct service delivery (Laprie, 1992).

A failure that can cause item loss (USA DoD, MIL-Std-721C, 1992).

Cause-Consequence Analysis

A method for illustrating the possible outcomes arising from the logical combination of selected input events or states (Jones, 1992).

Cause, Failure

See Failure Cause.

Cause Failure, Common

See Common Cause Failure.

Cause Fault, Common

See Common Cause Fault.

Cause, Root

See Root Cause.

Caution

Advisory in a software user document that performing some action may lead to consequences that are unwanted or undefined (IEEE Std-1063, 1988).

CCF

See Common Cause Failure.

CDR

See Critical Design Review.

Cell

See Process Cell (ISA, S88.01, 1995).

Certification

Procedure by which a third party gives written assurance that a product, process or service conforms to specified requirements (ISO/IEC, Guide 2, 1996).

The procedure and action by a duly authorised body of determining, verifying, and attesting in writing to the qualifications of personnel processes, procedures, or items in accordance with applicable requirements (USA DoD, MIL-Std-109C, 1994).

Acceptance by the certification authority that a process, product or demonstration satisfies a certification requirement (RTCA, DO178B, 1992).

The issue of a formal statement confirming the results of an evaluation, and that the evaluation criteria used were correctly applied (UK DTI, ITSEC, 1991).

The authoritative act of documenting compliance with requirements (BSI, BS 4778-2, 1991).

(1) A written guarantee that a system or component complies with its specified requirements and is acceptable for operational use. (2) A formal demonstration that a system or component complies with its specified requirements and is acceptable for operational use. (3) The process of confirming that a system or component complies with its specified requirements and is acceptable for operational use (IEEE Std-610.12, 1991).

The verification of a statement of truth by an independent officially recognised body or organisation (IEC 902, 1987).

Certification Body

Body that conducts certification (ISO/IEC, Guide 2, 1996).

An impartial body, governmental or non-governmental, possessing the necessary competence and integrity to operate a certification system, and on which the interests of parties concerned with the functioning of the system are represented (BSI, BS 4778-2, 1991).

Certification, Security

See Security Certification.

Certification System

A system having its own rules of procedure and management for carrying out certification (BSI, BS 4778-2, 1991).

Change, Engineering

See Engineering Change.

Change Notice, Specification

See Specification Change Notice.

Change, Software

See Software Change.

Channel

Element or a group of elements that independently perform(s) a function (IEC 61508-4, 1998).

Chargeable

Within the responsibility of a given organisational entity (applied to terms such as failures, maintenance time etc.) (USA DoD, MIL-Std-721C, 1992).

Chart, Sequential Function

See Sequential Function Chart.

Check

The test carried out before the system is put into commission (Hölscher & Rader, 1986).

Checking, Desk

See Desk Checking.

Checklist

A method for hazard identification by comparison with experience in the form of a list of failure modes and hazardous situations (Jones, 1992).

Checkout

Tests or observations of an item to determine its condition or status (USA DoD, MIL-Std-721C, 1992).

Testing conducted in the operational or support environment to ensure that a software product performs as required after installation (IEEE Std-610.12, 1991).

Checkout Phase, Installation and

See Installation and Checkout Phase.

Checkout Time

That element of maintenance time during which performance of an item is verified to be a specified condition (USA DoD, MIL-Std-721C, 1992).

Checkpoint

A point in a program, suitable for interrupting the execution of this program, at which a sequence of instructions is inserted to record the status and the results, to inspect them, and to restart (ISO/IEC DIS 2382-7, 1996).

A point in a computer program at which program state, status, or results are checked or recorded (IEEE Std–610.12, 1991).

CI

See Configuration Item.

Circuit Breaker

A device designed to open and close a circuit by non-automatic means, and also to open the circuit automatically on a predetermined over-current without damage to itself when properly applied within its designed rating (ASSE, 1988).

Circuit, Logic

See Logic Circuit.

Clock Time

The elapsed time from start to end of program execution, including wait time, on a running computer (Musa et al., 1987).

CM

See Configuration Management.

CMF

See Common Mode Failure.

Code

A piece of program text expressed in a programming language or in a form produced by an assembler, compiler, or other translator (ISO/IEC DIS 2382–7, 1996).

(1) In software engineering, computer instructions and data definitions expressed in a programming language or in a form output by an assembler, compiler, or other translator. (2) To express a computer

program in a programming language. (3) A character or bit pattern that is assigned a particular meaning (IEEE Std–610.12, 1991).

Any set of characters representing program instructions or data in any computer language (Smith & Wood, 1989).

Code, Deactivated

See Deactivated Code.

Code, Dead

See Dead Code.

Code, Inherited

See Inherited Code.

Code Inspection

See Inspection (IEEE Std–610.12, 1991).

Code, Object

See Object Code.

Code Review

A meeting at which software code is presented to project personnel, managers, users, customers, or other interested parties for comment or approval (IEEE Std–610.12, 1991).

Code, Source

See Source Code.

Code Template

A standard piece of proven code, for a particular function, which is used repetitively (Smith & Wood, 1989).

Coding

The transforming of logic and data from design specifications into a programming language (IEEE Std-990, 1987).

Coincident Error

Error produced on the same input (Laprie, 1992).

Cold Restart

Restart of a system and its application program after all dynamic data (variables such as I/O image, internal registers, timers, counters, etc., and program contexts) are reset to a predetermined state. A cold restart may be automatic (e.g. after a power failure, a loss of information in the dynamic portion(s) of the memory(ies) etc.) or manual (e.g. push-button reset, etc.) (IEC 1131-1, 1992).

Cold Stand-by

A configuration in which a redundant functional unit can be brought into service with some delay should the primary functional unit fail (ISO 2382-14, 1997).

Combinator

A physical structure or function used to combine a number of inputs to produce a single output. The output can also be affected by failures induced by the combinator itself (UK MoD, Def Stan 00-56, 1996).

Commission, Error of

See Error of Commission.

Commissioning

A collective term for the activities undertaken to prepare a system or product prior to demonstrating that it meets its specified requirements (CENELEC, prEN50126, 1998).

Commissioning Test

A test on a device or equipment carried out on site, to prove the correctness of installation and operation (IEC 902, 1987).

Common Cause Failure

A failure which is the result of an event(s) which causes a coincidence of failure states of two or more components leading to a system failing to perform its required operation (CENELEC, prEN50126, 1998).

Failures of multiple items occurring from a single cause which is common to all of them (ESA, ECSS-P-001A, 1997).

The result of a common cause fault (ISA, S84.01, 1996).

A single condition (e.g. corrosive environment) that slowly degrades multiple systems in ways that may eventually lead to failures (CCPS, 1993).

The failure of more than one component, item or system due to the same cause (Jones, 1992).

Failures arising from the same cause and hence unlikely to be independent of each other (Musa et al., 1987).

Common Cause Fault

A single source that will cause failure in multiple elements of a system. The single source may be either internal or external to the system (ISA, S84.01, 1996).

Common Failure Mode

Failures of distinct components caused by an initiating event (NATO, ARMP-7, 1996).

Common Mode Failure

Failures of multiple similar items that fail in the same mode (ESA, ECSS-P-001A, 1997).

Failure of apparently independent components or communication links due to an initiating event that affects them all (UK MoD, Def Stan 00–56, 1996).

A single failure that directly affects multiple systems in ways that are not consequences of each other (CCPS, 1993).

The failure of components in the same manner (Jones, 1992).

Failure resulting from related faults (Laprie, 1992).

Failure of items characterised by the same fault mode. (Note: common mode failures should not be confused with common cause failures as the common mode failures may result from different causes.) (BSI, BS 4778–3.2, 1991; IEC 50–191, 1990).

The result of an event which, because of dependencies, causes coincidence of failure states in two or more components (excluding secondary failures caused by the effects of a primary failure). Synonym to Common Cause Failure (IEC 812, 1985).

Common Mode Fault

Fault common to items which are intended to be independent (CENELEC, ENV50129, 1998).

Faults of multiple items which exhibit the same fault mode (ESA, ECSS-P–001A, 1997).

Communication, Internal

See Internal Communication.

Compatibility

Suitability of products, processes or services for use together under specific conditions to fulfil relevant requirements without causing unacceptable interactions (ISO/IEC, Guide 2, 1996).

Ability of entities to be used together under specific conditions to fulfil relevant requirements (ISO 8402, 1994).

The capability of a functional unit to meet the requirements of a specified interface without appreciable modification (ISO 2382-1, 1993).

The capability of a functional unit to meet the requirements of a specified interface (IEC 902, 1987).

Compatibility, Electromagnetic

See Electromagnetic Compatibility.

Compensation, Error

See Error Compensation.

Compile

To translate all or part of a program expressed in a high-level language into a program expressed in an intermediate language, an assembly language, or a machine language (ISO/IEC DIS 2382-7, 1996).

Complete

Information is complete when it includes necessary, relevant requirements and/or descriptive material, responses are defined for the range of valid input data, figures are labelled, and terms and units of measure are defined (RTCA, DO178B, 1992).

Complexity

The degree to which a system or component has a design or implementation that is difficult to understand and verify (UK MoD, Def Stan 00-55, 1997).

(1) The degree to which a system or component has a design or implementation that is difficult to understand and verify. (2) Pertaining to any of a set of structure-based metrics that measure the attribute in (1) (IEEE Std-610.12, 1991).

Complexity Metrix

A quantitative measure of the complexity of program structure, sometimes used to locate overcomplicated program modules (SSP 84, 1996; SSCP 83, 1995).

Compliance

A demonstration that a characteristic or property of a product satisfies the stated requirements (CENELEC, prEN50126, 1998).

An affirmative indication or judgement that the supplier of a product or service has met the requirements of the relevant specifications, contract or regulation; also the state of meeting the requirements (USA DoD, MIL-Std-109C, 1994).

Component

See Element (CENELEC, ENV50129, 1998).

A discrete structure, such as an assembly, within the total system, considered at a particular level of analysis (UK MoD, Def Stan 00-58, 1996; UK MoD, Def Stan 00-56, 1996).

Any device, unit, or application program which is a part of an automated system. However, its use in this standard is limited to manufacturer's catalogued parts of a Programmable Controller System (IEC 1131-1, 1992).

Item on the highest level considered in the analysis (IEC 300-3-1, 1991).

One of the parts that make up a system. A component may be hardware or software and may be subdivided into other components (IEEE Std-610.12, 1991).

An identifiable and self-contained portion of a Target of Evaluation (UK DTI, ITSEC, 1991).

Component, Computer Software

See Computer Software Component.

Component, Self-Checking

See Self-Checking Component.

Component, Software

See Software Component.

Component Testing

Testing of individual hardware or software components or groups of related components (IEEE Std–610.12, 1991).

Compression Factor, Testing

See Testing Compression Factor.

Computer

A functional unit that can perform substantial computations, including numerous arithmetic operations and logic operations without human intervention (ISO 2382–1, 1993).

An automatic data processing device (electrical or electronic) with capabilities of speed, accuracy, and memory that enable it to perform substantial and complex programmed computations without human intervention (ASSE, 1988).

Computer Architecture

The logical structure and functional characteristics of a computer, including the interrelationships among its hardware and software components (ISO 2382–1, 1993).

Computer Hardware

Devices capable of accepting and storing computer data, executing a systematic sequence of operations on computer data, or producing control outputs. Such devices can perform substantial interpretation, computation, communication, control, or other logical functions (USA DoD, MIL-Std–498, 1994; USA DoD, MIL-Hdbk–272A, 1993).

Computer Interface, Human

See Human Computer Interface.

Computer, Process

See Process Computer.

Computer Resources

The totality of computer hardware, software, personnel, documentation, supplies, and services applied to a given effort (USA DoD, MIL-Hdbk-347, 1990; USA DoD, MIL-Std-2167A, 1988).

Computer Software Component

A functionally or logically distinct part of a computer software configuration item, typically an aggregate of two or more software units (IEEE Std-610.12, 1991).

A distinct part of a computer software configuration item. CSCs may be further decomposed into other CSCs and Computer Software Units (USA DoD, MIL-Std-2167A, 1988).

Computer Software Documentation

Technical data or information, including computer listings and printouts, which documents the requirements, design, or details of computer software, explains the capabilities and limitations of the software, or provides operating instructions for using or supporting computer use during the software's operational life (USA DoD, MIL-Hdbk-347, 1990; USA DoD, MIL-Std-2167A, 1988).

Computer Software Unit

An element specified in the design of a Computer Software Component that is separately testable (USA DoD, MIL-Std-2167A, 1988).

Computer System

A system containing one or more computers and associated software (IEEE Std-610.12, 1991).

Computer System, Target

See Target Computer System.

Computer Utilisation

Fraction of clock time the processor is executing (Musa et al., 1987).

Concept Phase

The period of time in the software life cycle during which the user needs are described and evaluated through documentation (IEEE Std-1012, 1987; IEEE Std-1002, 1987).

Concession

A written authorisation to accept a configuration item, which, during production or after having been submitted for inspection, is found to depart from specified requirements, but nevertheless is considered suitable for use 'as is' or after rework by an approved method (UK MoD, Def Stan 05-57, 1993).

Waiver. Written authorisation to use or release a quantity of material, components or stores already produced but which do not conform to the specified requirements (BSI, BS 4778-1, 1991).

Concurrent

Consisting of components that function at the same time (Musa et al., 1987).

Condition Monitoring

The continuous or periodic measurement and interpretation of data to indicate the degraded condition of an item and the need for maintenance (BSI, BS 4778-3.1, 1991).

Conditional Jump

A jump that takes place only when the instruction that specifies it is executed and specified conditions are satisfied (IEC 1131–1, 1992).

Confidentiality

The prevention of the unauthorised disclosure of information (UK DTI, ITSEC, 1991).

Configuration

The structuring and interconnection of the hardware and software of a system for its intended application (CENELEC, ENV50129, 1998).

The functional and physical characteristics of material as described in technical documents and later achieved in a product (UK MoD, Def Stan 00–49, 1996; UK MoD, Def Stan 05–57, 1993; UK MoD, Def Stan 00–41, 1993).

All those components that make up the hierarchy of components for the control, protection and monitoring of the system (UK MoD, Def Stan 00–56, 1996).

The functional and physical characteristics of hardware, firmware, software or a combination thereof as set forth in technical documentation and achieved in a product (USA DoD, MIL-Std–109C, 1994; USA DoD, MIL-Hdbk–347, 1990; USA DoD, MIL-Std–480B, 1988).

(1) Interconnected equipment forming a system. (2) Application programming using vendor-embedded software (CCPS, 1993).

(1) The arrangement of a computer system or component as defined by the number, nature, and interconnections of its constituent parts. (2) In configuration management, the functional and physical characteristics of hardware or software as set forth in technical documentation or achieved in a product (IEEE Std–610.12, 1991).

The selection of one of the sets of possible combinations of features of a Target of Evaluation (UK DTI, ITSEC, 1991).

The specific arrangement of the programmable electronics within a PES and the combination of PES and non-PES safety-related systems (HSE, 1991).

(1) (Hardware.) A step in system design: selecting modules, assigning their locations and defining their interconnection. (2) (Software.) In a system or a device in which logic modules are pre-loaded, selecting required modules, assigning them to appropriate logical locations (e.g. control loop, display point, etc.) and linking them, mostly by keyboard operations. Contrasted with programming (IEC 902, 1987).

Configuration Audit

Checking an item for compliance with its configuration documentation (UK MoD, Def Stan 05–57, 1993).

Configuration Baseline

Configuration of a product, formally established at a specific point in time, which serves as a reference for other activities (ESA, ECSS-P-001A, 1997).

Configuration documentation formally designated and applicable at a specific point in the configuration item life cycle. The configuration baseline, plus the duly approved changes thereto, is defined in the current configuration identification (UK MoD, Def Stan 05–57, 1993).

Configuration Control

Activities comprising the control of changes to a configuration item after formal establishment of its configuration documents (ESA, ECSS-P-001A, 1997).

The techniques and processes used to ensure that only the correct versions of components are used in the design, development, production and in-service operation of the system (UK MoD, Def Stan 00–55, 1997).

The establishment of an agreed build standard for an item and the procedure for controlling change to that standard, in order that it may be defined at any time (NATO, ARMP-7, 1996).

The systematic proposal, justification, evaluation, co-ordination, approval or disapproval of proposed changes and the implementation of all approved changes, in the configuration of a CI after establishment of the configuration baseline(s) of the CI (USA DoD, MIL-Std-109C, 1994).

The systematic evaluation, co-ordination, approval or disapproval and dissemination of all proposed changes to a configuration item and its configuration documentation after formal establishment of its configuration baseline and verifying the implementation of all approved changes (UK MoD, Def Stan 05-57, 1993).

A system of controls imposed on changing controlled objects produced during the development, production and maintenance processes for a Target of Evaluation (UK DTI, ITSEC, 1991).

An element of configuration management, consisting of the evaluation, co-ordination, approval or disapproval, and implementation of changes to configuration items after formal establishment of their configuration identification (IEEE Std-610.12, 1991).

Configuration Documentation

The current approved technical documentation for a configuration item (UK MoD, Def Stan 05-57, 1993).

Configuration, Hardware

See Hardware Configuration.

Configuration Identification

(1) The process of designating the configuration items in a system and recording their characteristics. (2) The approved documentation that defines a configuration item (RTCA, DO178B, 1992).

(1) An element of configuration management, consisting of selecting the configuration items for a system and recording their functional and physical characteristics in technical documentation. (2) The current approved technical documentation for a configuration item as

set forth in specifications, drawings, associated lists, and documents reference therein (IEEE Std–610.12, 1991).

Configuration Item

Aggregation of hardware, software, processed materials, services or any of its discrete portions, that is designated for configuration management and treated as a single entity in the configuration management process (ESA, ECSS-P–001A, 1997).

An item that may be stored within the configuration management system (UK MoD, Def Stan 00–55, 1997).

An aggregation of hardware or software that satisfies an end use function and is designated by the Government for separate configuration management (USA DoD, MIL-Std–109C, 1994).

An item designated for configuration management (UK MoD, Def Stan 05–57, 1993).

(1) One or more hardware or software components treated as a unit for configuration management purposes. (2) The approved documentation that defines a configuration item (RTCA, DO178B, 1992).

An aggregation of hardware, software, or both, that is designated for configuration management and treated as a single entity in the configuration management process (IEEE Std–610.12, 1991).

Configuration Item, Software

See Software Configuration Item.

Configuration Management

A discipline applying technical and administrative direction and surveillance to identify and record the functional and physical characteristics of a configuration item, control changes to those characteristics, and record and report change control processing and implementation status (CENELEC, prEN50126, 1998; RTCA, DO178B, 1992; IEEE Std–610.12, 1991).

Technical and administrative procedures to: (1) identify and document functional and physical characteristics of any part of the system; (2) control changes to these, and their interfaces; (3) record and report change, its process and implementation (UK MoD, Def Stan 00-55, 1997).

The engineering and administrative activities which include configuration identification, configuration control, configuration accounting, configuration audit and the planning and surveillance of these activities (UK MoD, Def Stan 05-57, 1993).

The process of identifying and defining the configuration items of a system, controlling the release and change of these items throughout the software life cycle, recording and reporting the status of configuration items and change requests and verifying the completeness and correctness of configuration items (RTCA, DO178B, 1992).

The process of identifying configuration items, controlling changes, and maintaining the integrity and traceability of the configuration (CSA, CAN/CSA-Q396.1.1, 1989).

Configuring

Composition of a controlling system by selecting functional or modular units out of a given set and defining their interconnections (IEC 50-351, 1998).

Conformance Test

Tests that are specifically made to demonstrate compliance with relevant standards (IEC 902, 1987).

Conformance Testing

Testing whose purpose is checking whether the system satisfies its specification (Laprie, 1992).

Conformity

The fulfilling by an item or service of specification requirements (USA DoD, MIL-Std-109C, 1994).

Fulfilment of specified requirements (ISO 8402, 1994).

Consistency

The degree of uniformity, standardisation, and freedom from contradiction among the documents of a system or component (IEEE Std-610.12, 1991).

Consistent Failure

Failure perceived similarly by all system users (Laprie, 1992).

Constant Failure Period

The period of time in the life cycle of a system or component during which hardware failures occur at an approximately uniform rate (IEEE Std-610.12, 1991).

Consumer's Risk

The probability of accepting equipment with a true MTBF equal to the lower test MTBF. The probability of accepting equipment with a true MTBF less than the lower test MTBF will be less than the consumer's risk (NATO, ARMP-7, 1996).

The probability of an unacceptable batch being accepted due to a favourable sample (Smith, 1981).

Contingency Plan

A document that sets forth an organised, planned, and co-ordinated course of action to be followed in case of a fire, explosion, or release of hazardous waste or constituents thereof which could threaten human health or the environment (ASSE, 1988).

Continuous Mode

Safety Instrumented System designed to attain appropriate probability of a dangerous failure per year (e.g. Avionics) (ISA, S84.01, 1996).

Control

The activity to detect differences between an actual and a planned result/process, and to cause changes in a process or a product which reduce the detected differences to a defined level (UK MoD, Def Stan 05-95, 1995).

Purposeful action on or in a system, to meet specified objectives (IEC 902, 1987).

Control Algorithm

A mathematical representation of the control action to be performed (IEC 902, 1987).

Control, Basic

See Basic Control.

Control, Batch

See Batch Control.

Control Computer, Supervisory

See Supervisory Control Computer.

Control Coupling

The manner or degree by which one software component influences the execution of another software component (RTCA, DO178B, 1992).

Control Element, Final

See Final Control Element.

Control, Equipment

See Equipment Control.

Control, Feedback

See Feedback Control.

Control Module

The lowest level grouping of equipment in the physical model that can carry out basic control (ISA, S88.01, 1995).

Control, Open Loop

See Open Loop Control.

Control Program

A computer program designed to schedule and to supervise the execution of programs (RTCA, DO178B, 1992).

Control Recipe

A type of recipe which, through its execution, defines the manufacture of a single batch of a specific product (ISA, S88.01, 1995).

Control, Risk

See Risk Control.

Control System

System constituted by a controlled system, its controlling system, and the associated transducers (IEC 50-351, 1998).

A system which responds to input signals from the plant and/or from an operator and causes the plant to operate in the desired manner (HSE, 1991).

A system in which deliberate guidance or manipulation of one or more variables is used to achieve a prescribed state (IEC 902, 1987).

Control System, Basic Process

See Basic Process Control System.

Control System, Logic

See Logic Control System.

Control System, Process

See Process Control System.

Control System, Sequential

See Sequential Control System.

Controller, Programmable

See Programmable Controller.

Controller, Programmable Logic

See Programmable Logic Controller.

Controller System, Programmable

See Programmable Controller System.

Controls

In general, measures including devices, to regulate a machine, apparatus, system, or action within prescribed limits or standards of safety and operational effectiveness (ASSE, 1988).

Controls, Two-Hand

See Two-Hand Controls.

Conventional Safety Integrity

The level of safety integrity which has been achieved, in situations similar to those under consideration, by safety-related conventional systems which have traditionally been accepted as good engineering practice (HSE, 1991).

Conventional System

A non-programmable system or feature (HSE, 1991).

Correct Service

Service delivered in compliance with the system specification (Laprie, 1992).

Correction, Software

See Software Correction.

Corrective Action

Action taken to eliminate the causes of an existing nonconformity, defect or other undesirable situation in order to prevent recurrence (ESA, ECSS-P-001A, 1997; NATO, ARMP-7, 1996; ISO 8402, 1994).

A documented design, process, procedure, or materials change implemented and validated to correct the cause of failure or design deficiency (USA DoD, MIL-Std-721C, 1992).

Corrective Maintenance

The maintenance carried out after fault recognition and intended to put a product into a state in which it can perform a required function (CENELEC, prEN50126, 1998).

Maintenance carried out after occurrence of a failure, or detection of a fault, in order to restore a functional unit to a state in which it can perform a required function (ISO 2382-14, 1997).

Repair. The actions performed, as a result of failure, to restore an item to a specified condition (USA DoD, MIL-Hdbk-781A, 1996).

Maintenance actions carried out to restore a defective item to a specified condition (NATO, ARMP-7, 1996).

Preservation or improvement during its operational life of a system's ability to deliver a service complying with the specification. Fault removal during the operational life of a system (Laprie, 1992).

Reactive modification of a software product performed after delivery to correct discovered faults (IEEE Std-1219, 1992).

All actions performed as a result of failure, to restore an item to a specified condition. Corrective maintenance can include any of the following steps: localisation, isolation, disassembly, interchange, re-assembly, alignment and checkout (USA DoD, MIL-Std-721C, 1992).

Maintenance performed to correct faults in hardware or software (IEEE Std-610.12, 1991).

The actions associated with repair time (Smith, 1981).

Correctness

A property of a representation of a Target of Evaluation such that it accurately reflects the stated security target for that system or product (UK DTI, ITSEC, 1991).

(1) The degree to which a system or component is free from faults in its specification, design, and implementation. (2) The degree to which software, documentation, or other items meet specified requirements. (3) The degree to which software, documentation, or other items meet user needs and expectations, whether specified or not (IEEE Std-610.12, 1991).

The correctness of a unit is the conformity of the realised units with its specification (VDE 0801, 1990).

The correspondence between the realised and the specified function of some function entity (Hölscher & Rader, 1986).

Correctness, Proof of

See Proof of Correctness.

Correctness Proving

A formal mathematical demonstration that the semantics of a program is consistent with the specifications of that program (ISO/IEC DIS 2382-7, 1996).

Correctness, Total

See Total Correctness.

Corruption, Data

See Data Corruption.

Cost, Appraisal

See Appraisal Cost.

Cost, Prevention

See Prevention Cost.

Cost Risk

The degree of uncertainty associated with system acquisition life cycle budgets and outlays that may negatively impact the program (AFSC/AFLC 00-45, 1988).

COTS

Commercial off the Shelf.

Coupling

Interconnection or interdependence of different modules (ISO/IEC DIS 2382-7, 1996).

A measure of data connectivity between a module and the program in which it is embedded (USA DoD, MIL-Hdbk-272A, 1993).

The mechanism by which an interference source produces interference in a signal circuit (IEC 1131-1, 1992).

The manner and degree of interdependence between software modules (IEEE Std-610.12, 1991).

Coupling, Control

See Control Coupling.

Coupling, Data

See Data Coupling.

Coupling, Pathological

See Pathological Coupling.

Coverage

See Diagnostic Coverage (ISA, S84.01, 1996).

Measure of the representative of the situations to which a system is submitted during its validation compared to the actual situations it will be confronted with during its operational life (Laprie, 1992).

Sum of probabilities of occurrence of input states tested (Musa et al., 1987).

Coverage Analysis

The process of determining the degree to which a proposed software verification process activity satisfies its objective (RTCA, DO178B, 1992).

Coverage, Decision

See Decision Coverage.

Coverage, Diagnostic

See Diagnostic Coverage.

Coverage, Fault

See Fault Coverage.

Coverage, Repair

See Repair Coverage.

Coverage, Statement

See Statement Coverage.

Coverage, Test

See Test Coverage.

Covert

See Unrevealed (IEC 61508-4, 1998).

Covert Fault

Faults that can be classified as hidden, concealed, undetected, unrevealed, latent, etc. (ISA, S84.01, 1996).

CPC

Computer Program Component.

CPCI

Computer Program Configuration Item.

CPM

Critical Path Method.

CPQRA

Chemical Process Quantitative Risk Assessment.

CPU

Central Processing Unit.

Crash Failure

Persistent omission failure (Laprie, 1992).

CRC

Cyclic Redundancy Check.

Credible

Believed to have a probability of occurrence that can be expressed in meaningful numerical terms (UK MoD, Def Stan 00-56, 1996).

Criteria, Pass/Fail

See Pass/Fail Criteria.

Criteria, Safety Integrity

See Safety Integrity Criteria.

Criteria, Test

See Test Criteria.

Critical Anomaly

An anomaly that must be resolved before the V&V effort proceeds to the next life cycle phase (IEEE Std-1012, 1987).

Critical Defect

A defect that would result in hazardous or unsafe conditions for individuals using, maintaining, or depending upon the product; or a defect that judgement and experience indicate is likely to prevent performance of the tactical function of a major end item or major part thereof (USA DoD, MIL-Std-109C, 1994).

A defect that analysis, judgement and experience indicate is likely to result in hazardous or unsafe conditions (BSI, BS 4778-2, 1991).

Critical Design Review

(1) A review conducted to verify that the detailed design of one or more configuration items satisfy specified requirements; to establish the compatibility among the configuration items and other items of equipment, facilities, software, and personnel; to assess risk areas for each configuration item; and, as applicable, to assess the results of producibility analyses, review preliminary hardware product specifications, evaluate preliminary test planning, and evaluate the adequacy of preliminary operation and support documents. (2) A review as in (1) of any hardware or software component (IEEE Std-610.12, 1991).

Critical Failure

A failure that could result in injury to persons or that prevents an item from performing an essential mission (NATO, ARMP-7, 1996).

A failure, or combination of failures, that prevents an item from performing a specified mission (USA DoD, MIL-Std-721C, 1992).

A failure which is both sudden and causes cessation of one or more fundamental functions. This failure requires immediate corrective action in order to return the item to a satisfactory condition (OREDA, 1992).

A failure which is assessed as likely to result in injury to persons, significant material change or other unacceptable consequences (BSI, BS 4778-3.2, 1991; IEC 50-191, 1990).

Critical Fault

A fault which is assessed as likely to result in injury to persons, significant material damage, or other unacceptable consequences (BSI, BS 4778–3.2, 1991; IEC 50–191, 1990).

Critical Item

An item whose failure could result in a critical failure or that requires special effort during development/production (NATO, ARMP–7, 1996).

In configuration management, an item within a configuration item that, because of special engineering or logistic considerations, requires an approved specification to establish technical or inventory control at the component level (IEEE Std–610.12, 1991).

Critical, Safety

See Safety-Critical.

Critical Software

Software whose failure could have an impact on safety, or could cause large financial or social loss (IEEE Std–610.12, 1991; IEEE Std–730, 1989; IEEE Std–1012, 1987).

Criticality

Highest severity of failure modes (Laprie, 1992).

A relative measure of the consequence of a failure mode and its frequency of occurrences (USA DoD, MIL-Std–721C, 1992).

Numerical index of the severity of an effect combined with the probability or expected frequency of its occurrence (BSI, BS 4778–3.1, 1991).

The degree of impact that a requirement, module, error, fault, failure, or other item has on the development or operation of a system (IEEE Std–610.12, 1991).

Cross-Coupling

The degree of interdependence between program modules (SSP 84, 1996; SSCP 83, 1995).

Cross-Reference List

A list showing all the locations in a PC program at which a selected list of data objects is operated upon (IEC 1131-1, 1992).

A list that identifies each of the variables, labels, and other identifiers in a computer program and indicates which statements in the program define, set, or use each one (IEEE Std-610.12, 1991).

CSC

See Computer Software Component.

CSU

See Computer Software Unit.

Curative Maintenance

Corrective maintenance aimed at removing faults which have produced errors and which have been reported (Laprie, 1992).

Customer

Recipient of a product provided by the supplier (IEC 300-3-4, 1996; ISO 8402, 1994).

The person, or persons, for whom the product is intended, and usually (but not necessarily) who decides the requirements (IEEE Std-1219, 1992).

The person or organisation that purchases a Target of Evaluation (UK DTI, ITSEC, 1991).

Cycle, Software Development

See Software Development Cycle.

Cycle, System Development

See System Development Cycle.

D

Damage

Loss in value, usefulness, etc. to property or things. Harm causing any material loss (ASSE, 1988).

Danger

The likelihood of the hazard leading to an accident (Leveson, 1995).

The complement of safety (Hölscher & Rader, 1986).

Dangerous Failure

Failure which has the potential to put the safety-related system in a hazardous or fail-to-function state (IEC 61508-4, 1998).

Failure of at least one of the functions which is necessary for safety in a unit carrying out some task (Hölscher & Rader, 1986).

Data

A representation of facts, concepts or instructions in a formalised manner suitable for communication, interpretation or 'processing' by human or by 'automatic' means (ISO 2382-1, 1993; IEEE Std-610.12, 1991).

Characters or continuous functions representing information due to known or supposed arrangement (IEC 1131-1, 1992; IEC 902, 1987).

Data Corruption

The introduction of an error by reason of some alteration of the software already resident in the system. This could arise from electrical, magnetic or ionising interference or from incorrect processing of a portion of the software (Smith & Wood, 1989).

Data Coupling

The dependence of a software component on data not exclusively under the control of that software component (RTCA, DO178B, 1992).

Data Exception

An exception that occurs when a program attempts to use or access data incorrectly (IEEE Std-610.12, 1991).

Data Protection

The implementation of appropriate administrative, technical or physical means to guard against unauthorised intentional or accidental disclosure, modification, or destruction of data (ISO 2382-1, 1993).

Data-Sensitive Fault

A fault that causes a failure in response to some particular pattern of data (IEEE Std-610.12, 1991).

DCI

Diagnostic Check Interval.

DCS

Distributed Control System.

DCSC

Distributed Control System Controller.

DDC

Direct Digital Control.

Deactivated Code

Executable object code (or data) which by design is either (1) not intended to be executed (code) or used (data), e.g. a part of a previously developed software component, or (2) is only executed (code) or used (data) in certain configurations of the target computer environment, e.g. code that is enabled by a hardware pin selection or software programmed options (RTCA, DO178B, 1992).

Dead Code

Executable object code (or data) which, as a result of a design error cannot be executed (code) or used (data) in an operational configuration of the target computer environment and is not traceable to a system or software requirement. An exception is embedded identifiers (RTCA, DO178B, 1992).

Dead Time, Fractional

See Fractional Dead Time.

Deadlock

A situation in which data processing is suspended because two or more device or concurrent processes are each awaiting resources assigned to the other(s) or because of other mutual dependencies (ISO/IEC DIS 2382-7, 1996).

A situation in which computer processing is suspended because two or more devices or processes are each awaiting resources assigned to the others (IEEE Std-610.12, 1991).

Death

In relation to an accident, an injury which terminates fatally and is causally related to the accident (ASSE, 1988).

Debug

To detect, to locate, and to eliminate errors in programs (ISO/IEC DIS 2382-7, 1996).

To detect, locate, and eliminate errors in programs (ISO 2382-1, 1993).

To detect, locate, and correct faults in a computer program (IEEE Std-610.12, 1991).

Debugging

A process to detect and remedy inadequacies. Not to be confused with terms such as burn-in, fault isolation or screening (USA DoD, MIL-Std-721C, 1992).

Debugging Tool, Programming and

See Programming and Debugging Tool.

Decision Coverage

Every point of entry and exit in the program has been invoked at least once and every decision in the program has taken on all possible outcomes at least once (RTCA, DO178B, 1992).

Decommissioning

The permanent removal of a complete Safety Instrumented System from active service (ISA, S84.01, 1996).

De-Energise To Trip

SIS circuits where the outputs and devices are energised under normal operation. Removal of the source of power (e.g. electricity, air) causes a trip action (ISA, S84.01, 1996).

Defect

Any non-conformance of an item with specified requirements, or a condition which experience indicates could result in a non-conformance (NATO, ARMP-7, 1996).

Any non-conformance of the unit of product with specified requirements or any state or condition of non-conformance to requirements (USA DoD, MIL-Std-109C, 1994).

Non-fulfilment of an intended usage requirement or reasonable expectation, including one concerned with safety (ISO 8402, 1994).

The non-fulfilment of intended usage requirements (BSI, BS 4778-2, 1991; BSI, BS 4778-1, 1991; IEC 902, 1987).

A product anomaly. Examples include such things as (1) omissions and imperfections found during early life cycle phases and (2) faults contained in software sufficiently mature for test or operation (IEEE Std-982.1, 1988).

Inadmissible deviation from some property of a system or the non-fulfilment of some specified requirements by the value of some property (Hölscher & Rader, 1986).

Defect, Critical

See Critical Defect.

Defective

A unit of product which contains one or more defects (USA DoD, MIL-Std-109C, 1994).

Defensive Code

See Defensive Design (UK MoD, Def Stan 00-55, 1997).

Defensive Design

Design or code which prevents the propagation of errors, by incorporating checks even for circumstances which are believed to be infeasible (UK MoD, Def Stan 00–55, 1997).

Deficiency

(1) Conditions or characteristics in any item which are not in accordance with the item's current approved configuration documentation. (2) Inadequate (or erroneous) item configuration documentation which has resulted, or may result, in units of the item that do not meet the requirements for the item (USA DoD, MIL-Std–973, 1992).

Degradation

A gradual impairment in ability to perform (UK MoD, Def Stan 00–49, 1996; USA DoD, MIL-Std–721C, 1992).

Degradation Fault

A fault which reduces performance below a specified level, but which does not necessarily terminate the function (UK MoD, Def Stan 00–49, 1996).

Degradation, Graceful

See Graceful Degradation.

Degraded Failure

A failure which is gradual, partial, or both. Such a failure does not cease the fundamental functions, but compromises one or several functions. The function may be compromised by any combination of reduced, increased, or erratic outputs, In time, such a failure may develop into a critical failure (OREDA, 1992).

Delay Time

That element of down-time during which no maintenance is being accomplished on the item because of either supply or administrative delay (USA DoD, MIL-Std-721C, 1992).

The delay encountered by the signal as it passes through the input filter (from the field input terminal(s) to the input conversion circuit boundary) (IEC 1131-1, 1992).

Demand

A condition or event that requires the Safety Instrumented System to take appropriate action to prevent a hazardous event from occurring or mitigate the consequence of a hazardous event (ISA, S84.01, 1996).

A plant condition or event which requires the Instrumented Protective System to take action to prevent a hazard (Korteweg, 1994).

A condition or event which requires a protective system or device to take appropriate action to prevent or mitigate a hazard (CCPS, 1993).

A condition which requires a protective system to operate (Jones, 1992).

Demand Mode

Safety Instrumented System designed to attain appropriate probability of failure to perform its design function on demand (ISA, S84.01, 1996).

Demand, Probability of Failure on

See Probability of Failure on Demand.

Demand Rate

The frequency at which demands occur. Dimension (time^-1) (Korteweg, 1994).

Demonstrated

That which has been measured by the use of objective evidence gathered under specified conditions (USA DoD, MIL-Std-721C, 1992).

Demonstration

A dynamic analysis technique that relies on observation of system or component behaviour during execution, without need for postexecution analysis, to detect errors, violations of development standards, and other problems (IEEE Std-610.12, 1991).

Dependability

The collective term used to describe the availability performance and its influencing factors: reliability performance, maintainability performance and maintenance support performance (ESA, ECSS-P-001A, 1997; IEC 300-3-4, 1996; ISO 8402, 1994; BSI, BS 4778-3.2, 1991; IEC 50-191, 1990).

A measure of the degree to which an item is operable and capable of performing its required function at any (random) time during a specified mission profile, given item availability at the start of the mission (USA DoD, MIL-Std-109C, 1994; USA DoD, MIL-Std-721C, 1992).

The collective term used to describe the availability performance and its influencing factors: reliability performance, maintainability performance and maintenance support performance. (Note: Dependability is used only for general descriptions in non-quantitative terms) (ISO 9000-4, 1993; IEC 300-1, 1993).

Trustworthiness of a computer system such that reliance can justifiably be placed on the service it delivers (Laprie, 1992).

Dependability, Impairments to

See Impairments to Dependability.

Dependability Programme

The organisational structure, responsibilities, procedures, processes and resources used for managing dependability (ISO 9000–4, 1993; IEC 300–1, 1993).

Dependent Failure

Failure whose probability cannot be expressed as the simple product of the unconditional probabilities of the individual events which caused it (IEC 61508–4, 1998).

The failure of a set of events, the probability of which cannot be expressed as the simple product of the unconditional probabilities of the individual events (CENELEC, prEN50126, 1998).

Failure which is caused by the failure of an associated item(s). Not independent (USA DoD, MIL-Std–721C, 1992).

Derating

Process of designing a product such that its components operate at a significantly reduced level of stress to increase reliability (ESA, ECSS-P–001A, 1997).

Using an item in such a way that applied stresses are below rated values (NATO, ARMP–7, 1996; USA DoD, MIL-Std–721C, 1992).

The lowering of the rating of an item in one stress field to allow an increase in another stress field (USA DoD, MIL-Std–721C, 1992).

The use of components having a higher strength rating in order to reduce failure rate (Smith, 1981).

Derived Requirements

Additional requirements resulting from design or implementation decisions during the development process. Derived requirements are not directly traceable to higher level requirements; though derived requirements can influence higher level requirements (UK MoD, Def Stan 00–55, 1997).

Additional requirements resulting from the development processes, which are not directly traceable to higher-level requirements (RTCA, DO178B, 1992).

Description, Design

See Design Description.

Design

The prebuilt exercise of defining elements and their interconnection such that the product will meet its intended requirements (CENELEC, ENV50129, 1998).

(1) Set of information which defines the essential characteristics of a product. (2) The process used to generate the set of information describing the essential characteristics of a product (ESA, ECSS-P-001A, 1997).

(1) The process of defining the architecture, components, interfaces, and other characteristics of a system or component. (2) The result of the process in (1) (IEEE Std-610.12, 1991).

Design Approval

See Test Qualification (USA DoD, MIL-Std-721C, 1992).

Design, Defensive

See Defensive Design.

Design Description

A document that describes the design of a system or component (IEEE Std-610.12, 1991).

Design Diversity

An approach to the production of systems, involving the provision of identical services from separate designs and implementations (Laprie, 1992).

Design Fault

Human-made internal fault (Laprie, 1992).

Design for Verifiability

Methods and techniques when designing a system which facilitate its verification (Laprie, 1992).

Design, Formal

See Formal Design.

Design, Functional

See Functional Design.

Design Intent

The designer's required or specified behaviour, or range of behaviour, of an aspect of the design (usually an attribute) (UK MoD, Def Stan 00–58, 1996).

Design, Preliminary

See Preliminary Design.

Design Representation

A descriptive model of the design of a system or part of a system, presented according to a convention (UK MoD, Def Stan 00–58, 1996).

Design Review

A formal and independent examination of an existing or proposed design for the purpose of detection and remedy of deficiencies in the requirements and design which could affect such things as reliability performance, maintainability performance, maintenance support performance requirements, fitness for the purpose and the identification of potential improvements (UK MoD, Def Stan 00–49, 1996).

A formal, documented, comprehensive and systematic examination of a design to evaluate the design requirements and the capability of the design to meet these requirements and to identify problems and propose solutions (USA DoD, MIL-Std-109C, 1994; BSI, BS 4778-3.2, 1991; BSI, BS 4778-1, 1991; IEC 50-191, 1990).

A formal appraisal at appropriate stages of design, development, production and in-service to evaluate achievement of the reliability and maintenance requirements (UK MoD, Def Stan 00-41, 1993).

A design review is a formal, documented engineering management process that is used to subject a design to a systematic critical study. Its purpose is to establish that the design satisfies the specified requirement (UK MoD, Def Stan 05-57, 1993).

A process or meeting during which a system, hardware, or software design is presented to project personnel, managers, users, customers, or other interested parties for comment or approval (IEEE Std-610.12, 1991).

Design Review, Critical

See Critical Design Review.

Design Review, Formal

See Formal Design Review.

Design Review, Preliminary

See Preliminary Design Review.

Design Review, System

See System Design Review.

Design Specification

During the course of a design specification, the unit is a particular part of the system (e.g. the software), whereby individual design decisions have already been taken. These decisions are called design specifications (VDE 0801, 1990).

Design, Structured

See Structured Design.

Design, Test

See Test Design.

Desk Checking

A static analysis technique in which code listings, test results, or other documentation are visually examined, usually by the person who generated them, to identify errors, violations of development standards, or other problems (IEEE Std-610.12, 1991).

Detected

Revealed. Overt. In relation to hardware, detected by the diagnostic tests, proof tests, operator intervention (for example physical inspection and manual tests), or through normal operation (IEC 61508-4, 1998).

Detected Error

Error recognised as such by a detection algorithm or mechanism (Laprie, 1992).

Detected Fault

See Overt Fault (ISA, S84.01, 1996).

Detection, Error

See Error Detection.

Determinate Fault

For an item which produces a response as a result of an action, a fault for which the response is the same for all actions (BSI, BS 4778-3.2, 1991; IEC 50-191, 1990).

Deterministic

Possessing the property of having only one value at a given time (Musa et al., 1987).

Deterministic Testing

Form of testing where the test patterns are predetermined by a selective choice (Laprie, 1992).

Testing in which all selection of runs has been planned and prescribed in advance (Musa et al., 1987).

Developer

A person or organisation that performs development activities (including requirements analysis, design, testing through acceptance) during the software life cycle process (IEEE Std-1062, 1992).

The person or organisation that manufactures a Target of Evaluation (UK DTI, ITSEC, 1991).

Development

The process by which the capability to adequately implement a technology or design is established before manufacture (ESA, ECSS-P-001A, 1997).

The process by which the capability to adequately implement a technology or design is established before manufacture (ESA, ECSS-P-001A, 1997).

All activities that are carried out to create a software product (IEEE Std-1298, 1992; SA, AS 3563, 1991).

Development Baseline

The configuration documentation formally designated at the beginning of design and development prescribing: (1) the functional characteristics that are allocated from the functional baseline and/or from higher-level configuration items to lower level configuration items; (2) the tests required to demonstrate achievement of those functional characteristics; (3) the necessary interface characteristics with associated configuration items; and (4) design constraints (UK MoD, Def Stan 05-57, 1993).

Development Cycle, Software

See Software Development Cycle.

Development Cycle, System

See System Development Cycle.

Development Process

A set of tasks which, when carried out in a defined manner on identified inputs and outputs, constitute an identifiable element of the SRS development life cycle (UK MoD, Def Stan 00-55, 1997).

The set of phases and tasks whereby a Target of Evaluation is constructed, translating requirements into actual hardware and software (UK DTI, ITSEC, 1991).

Development, Software

See Software Development.

Development Testing

Formal or informal testing conducted during the development of a system or component, usually in the development environment by the developer (IEEE Std-610.12, 1991).

Developmental Item

An item to be modified, expanded, created, or that is in any stage of being created (USA DoD, MIL-Hdbk-1467, 1997).

Deviation

See Production Permit (ESA, ECSS-P-001A, 1997; UK MoD, Def Stan 05-57, 1993).

An excursion of the value of an attribute of the system from its design intent (UK MoD, Def Stan 00-58, 1996).

(1) A specific written authorisation, granted prior to the manufacture of an item, to depart from a particular requirement(s) of an item's current approved configuration documentation for a specific number of units or a specified period of time. (2) The difference or distance of an individual observation or data value from the centre point (often the mean) of the data set distribution (USA DoD, MIL-Std-109C, 1994).

(1) A departure from a specified requirement. (2) A written authorisation, granted prior to the manufacture of an item, to depart from a particular performance or design requirement for a specific number of units or a specific period of time (IEEE Std-610.12, 1991).

Device

Part of a Programmable Controller System (IEC 1131-1, 1992).

A mechanism or piece of equipment designed to serve a purpose or perform a function (IEEE Std-610.12, 1991).

An apparatus or hardware element for performing a prescribed function (IEC 902, 1987).

Device, Output

See Output Device.

Device, Programmable

See Programmable Device.

Devices, Field

See Field Devices.

DFD

Data Flow Diagram.

DFR

Decreasing Failure Rate.

Diagnosis, Failure

See Failure Diagnosis.

Diagnosis, Fault

See Fault Diagnosis.

Diagnostic

Pertaining to the detection and isolation of faults or failures (IEEE Std–610.12, 1991).

Diagnostic Coverage

Fractional decrease in the probability of dangerous hardware failure resulting from the operation of the automatic diagnostic tests (IEC 61508–4, 1998).

For Safety Instrumented System with active fault-detection capabilities, the ratio of detectable faults to the total number of faults (ISA, S84.01, 1996).

Diagnostic Function

The capability of the hardware to diagnose internal problems and indicate those errors as a status (IEC 902, 1987).

Diagnostic Program

A program that is designed to detect, locate and describe faults in equipment or errors in programs (ISO/IEC DIS 2382-7, 1996).

A troubleshooting program for identifying hardware or software malfunctions (CCPS, 1993).

Diagnostic Software

A program which assists in locating and explaining the causes of faults in software (Smith & Wood, 1989).

Diagnostics

The detection, isolation and analysis of faults and failures (NATO, ARMP-7, 1996).

DID

Design Input Documentation.

Digital Signal

A signal the information parameter of which may assume any value out of a set of discrete values which are represented by numbers (IEC 902, 1987).

Direct Metric

A metric applied during development or during operations that represents a software quality factor (IEEE Std-1061, 1992).

Discrepancy

Difference between actual value of output variable, resulting from a run, and the value prescribed by the requirements (Musa et al., 1987).

Discrimination Ratio

The ratio of passing through an acceptance test.

Disturbance

Impairment of a function. It is used, however, only for those impairments of function which originate from outside the system under consideration: a disturbance is an action, not a state (Hölscher & Rader, 1986).

Diverse

Use of different technologies, equipment or design methods to perform a common function with the intent to minimise common cause faults (ISA, S84.01, 1996).

Diverse Redundancy

See Diversity (IEEE Std–610.12, 1991).

Redundancy with means which are not the same. The mains which are not the same can e.g. be (1) other physical principles, (2) other ways of solving the same task, (3) other design procedures etc. (VDE 0801, 1990).

Diversity

Different means of performing a required function (IEC 61508–4, 1998).

The existence of different means of performing a required function, for example, other physical principles, other ways of solving the same problem (UK MoD, Def Stan 00–55, 1997).

A number of independent and different means to perform the same overall protective function (CCPS, 1993).

The performance of the same function by a number of independent and different means (Jones, 1992).

Dissimilar technical devices used for the purpose of achieving useful redundancy (Hölscher & Rader, 1986).

Diversity, Software

See Software Diversity.

Documentation

The written (or otherwise recorded) information about a Target of Evaluation required for an evaluation. This information may, but need not, be contained within a single document produced for the specified purpose (UK DTI, ITSEC, 1991).

(1) A collection of documents on a given subject. (2) Any written or pictorial information describing, defining, specifying, reporting, or certifying activities, requirements, procedures, or results. (3) The process of generating or revising a document. (4) The management of documents, including identification, acquisition, processing, storage, and dissemination (IEEE Std–610.12, 1991).

Documentation, Computer Software

See Computer Software Documentation.

Documentation, Configuration

See Configuration Documentation.

Domain

The part of the external world, including users and personnel, that affects and is affected by the system and may be harmed by an accident caused by it (UK MoD, Def Stan 00–56, 1996).

Dormant

See Not Operating (USA DoD, MIL-Std-721C, 1992).

Dormant Fault

Internal fault not activated by the computation process (Laprie, 1992).

Double Knock

Redundant 1 out of 2 logical AND function.

Down

Pertaining to a system or component that is not operational or has been taken out of service (IEEE Std-610.12, 1991).

Down-State

A state of an item characterised either by a fault, or by a possible inability to perform a required function during preventive maintenance (BSI, BS 4778-3.2, 1991; IEC 50-191, 1990).

Down-Time

The time interval during which a product is in a down-state (CENELEC, prEN50126, 1998; BSI, BS 4778-3.2, 1991; IEC 50-191, 1990).

That period of time during which an item is not in a condition to perform a required function. It is the sum of the total preventive maintenance (TPM) time, plus total corrective maintenance (TCM) time plus the total of the administration and logistics delay time (ALDT) spent waiting for parts, maintenance personnel or transportation (UK MoD, Def Stan 00-49, 1996).

That element of active time during which an item is not in condition to perform its required function (reduces availability and dependability) (USA DoD, MIL-Std-721C, 1992).

The total time during which the system is not in acceptable operating condition (Martz & Waller, 1982).

The time during which an item is not able to perform to specification. DT = SUM(realisation-, access-, diagnosis-, spares-, replace-, check-, align-time) = SUM(realisation-time, repair-time) (Smith, 1981).

Downing Event

The event which causes an item to become unavailable to initiate its mission (the transition from up-time to down-time) (USA DoD, MIL-Std–721C, 1992).

DRACAS

Data Reporting Analysis and Corrective Action System.

Driver, Test

See Test Driver.

Drivers, Risk

See Risk Drivers.

DT

See Down-Time.

DTE

Data Terminal Equipment.

Dump

(1) To record or display in a format that facilitates analysis, the contents of all or part of a storage device at a particular instant. (2) Data that have been dumped (ISO/IEC DIS 2382–7, 1996).

Dump, Post-Mortem

See Post-Mortem Dump.

Duplication

Method to achieve reliability. The use of additional, redundant, parts whose individual failure does not cause the overall product to fail (Smith, 1981).

Durability

The ability of a functional unit to perform a required function under given conditions of use and maintenance, until a limiting state is reached (ISO 2382-14, 1997; UK MoD, Def Stan 00-49, 1996).

The ability of an item or material to resist cracking (including stress corrosion and hydrogen induced cracking), corrosion, thermal shock and degradation, delamination, wear and the effects of foreign object damage for a specified period of time (NATO, ARMP-7, 1996).

A measure of useful life (a special case of reliability) (USA DoD, MIL-Std-721C, 1992).

DUT

Device Under Test.

Dynamic Analysis

The investigation of software properties by analysing execution of the code (SSCP 83, 1995).

Dynamic Error

An error that is dependent on the time-varying nature of an input (IEEE Std-610.12, 1991).

Dynamic Testing

Executing software in a controlled and systematic way, so as to demonstrate the presence of the required behaviour and the absence of unwanted behaviour (IEC 61508-4, 1998).

Dynamic Verification

Verification involving exercising the system (Laprie, 1992).

E

E

(1) *See* Electrical. (2) *See* Electronic.

Early Failure Period

The period of time in the life cycle of a system or component during which hardware failures occur at a decreasing rate as problems are detected and repaired (IEEE Std–610.12, 1991).

Early Life Failure

Premature Failure. Failures which occur in the early life of an item. The early life of an item is normally characterised as a period in which the failure rate of the item is considerably higher than that of the subsequent period. Such failures are usually amenable to removal by burn-in or environmental stress screening (NATO, ARMP–7, 1996).

ECP

See Engineering Change Proposal.

E/E/PE

Electrical/Electronic/Programmable Electronic.

Effect, Failure

See Failure Effect.

Effect, Side

See Side Effect.

Effective Availability

The proportion of the defined period during which the equipment is available for use without any performance limitations, assuming that there are no logistic delays (UK MoD, Def Stan 00–49, 1996).

Effectiveness

A property of a Target of Evaluation representing how well it provides security in the context of its actual or proposed operational use (UK DTI, ITSEC, 1991).

Effectiveness, System

See System Effectiveness.

Efficiency

The degree to which a system or component performs its designated functions with minimum consumption of resources (IEEE Std–610.12, 1991).

Electrical

When used in the context of E/E/PES, electrical refers to logic functions performed by electromechanical techniques, e.g. electromechanical relay, motor-driven timers, etc. In E/E/PES the first 'E' refers to electrical (ISA, S84.01, 1996).

Electrical Isolation

Separation of electrical circuits, signals, or data to preclude ambiguity, interference, or information perversion. This may be achieved through physical isolation or by any property which distinguishes one electrical signal from all others (USA DoD, MIL-Hdbk–272A, 1993).

Electromagnetic Compatibility

The ability of electronic equipment, subsystems, and systems to operate together in their intended operational environments without suffering or causing unacceptable degradation because of unintentional electromagnetic radiation or response (USA DoD, MIL-Hdbk-272A, 1993).

Electromagnetic Interference

Any electromagnetic disturbance which interrupts, operates, obstructs, or otherwise degrades or limits the effective performance of electronics/electrical equipment (USA DoD, MIL-Hdbk-272A, 1993).

Electronic

When used in the context of E/E/PES, electronic refers to logic functions performed by electronic techniques, e.g. solid state logic, solid state relay, etc. In E/E/PES the second 'E' refers to electronic (ISA, S84.01, 1996).

Electronic, Programmable

See Programmable Electronic.

Electronic System, Programmable

See Programmable Electronic System.

Electronics, Programmable

See Programmable Electronics.

Element

Component. A part of a product that has been determined to be a basic unit or building block. An element may be simple or complex (CENELEC, ENV50129, 1998).

The portion of a graphical language corresponding to an instruction in textual language. The operation part and the operand part as specified by appropriate combination of graphic and character symbols (IEC 1131-1, 1992).

Embedded Software

Software that is part of the system supplied by the vendor and is not accessible for modification by the end user. Embedded software is also referred to as firmware or system software (ISA, S84.01, 1996).

Emergency

Condition when potentially catastrophic or critical hazardous events have occurred, and immediate and preplanned safing action is required (ESA, ECSS-P-001A, 1997).

Emergency Alarm

A warning device, usually visual or auditory, which indicates the existence of an emergency situation requiring immediate action (ASSE, 1988).

Emergency Maintenance

Unscheduled corrective maintenance performed to keep a system operational (IEEE Std-1219, 1992).

Emergency Shutdown System

See Safety Instrumented System (ISA, S84.01, 1996).

A term that is being superseded by Safety Interlock System (CCPS, 1993).

Emergency Switch

A switch placed in a convenient position for cutting off the supply of electricity to a piece of equipment or to a building in case of emergency (ASSE, 1988).

EMI

See Electromagnetic Interference.

Emulation

The use of a data processing system to imitate another data processing system, so that the imitating system accepts the same data, executes the same programs, and achieves the same results as the imitated system (ISO 2382-1, 1993).

(1) A model that accepts the same inputs and produces the same outputs as a given system. (2) The process of developing or using a model as in (1) (IEEE Std-610.12, 1991).

A type of simulation whereby the simulator responds to all possible inputs as would the real item and generates all the corresponding outputs (Smith & Wood, 1989).

Emulator

A device, computer program, or system that accepts the same inputs and produces the same outputs as a given system (IEEE Std-610.12, 1991).

Encapsulation

A software development technique that consists of isolating a system function or a set of data and operations on that data within a module and providing precise specifications for the module (UK MoD, Def Stan 00-55, 1997).

A software development technique that consists of isolating a system function or a set of data and operations on those data within a module and providing precise specifications for the module (IEEE Std-610.12, 1991).

Endurance Test

A test carried out over a time interval to investigate how the properties of an item are affected by the application of stated stresses and by their time duration or repeated application (BSI, BS 4778-3.2, 1991; IEC 50-191, 1990).

Load test. Stress test. Extreme testing whereby the software is subjected to abnormal and illegal inputs and conditions and saturation levels which stress the program capabilities in terms of data volume and rate, processing time, utilisation of memory, etc. (Smith & Wood, 1989).

Energise To Trip

Safety Instrumented System circuits where the outputs and devices are de-energised under normal operation. Application of power (e.g. electricity, air) causes a trip action (ISA, S84.01, 1996).

Engineering Change

An alteration in the configuration of a configuration item and/or its configuration documentation after formal establishment of its configuration baseline (UK MoD, Def Stan 05-57, 1993).

A change to the current approved configuration documentation of a configuration item at any point in the life cycle of the item (USA DoD, MIL-Std-973, 1992).

Engineering Change Proposal

A proposed engineering change and the documentation by which the change is described, justified, and submitted to the government for approval or disapproval (USA DoD, MIL-Std-973, 1992).

Engineering Environment, Software

See Software Engineering Environment.

Engineering Release

An action whereby configuration documentation or an item is officially made available for its intended use (USA DoD, MIL-Std-973, 1992).

Engineering, Reverse

See Reverse Engineering.

Enhancement, Software

See Software Enhancement.

Entity, Equipment

See Equipment Entity.

Environment

Conditions in which an item exists or is operated (ESA, ECSS-P-001A, 1997).

A collection of hardware and software tools to support one or more phases of the software development cycle (ISO/IEC DIS 2382-7, 1996).

The other systems having interacted or interfered, interacting or interfering, or likely to interact or interfere with the considered system (Laprie, 1992).

The aggregate of all external and internal conditions (such as temperature, humidity, radiation, magnetic and electric fields, shock vibration, etc.) either natural or man made, or self-induced, that influences the form, performance, reliability or survival of the item (USA DoD, MIL-Std-721C, 1992).

Environment, Operational

See Operational Environment.

Environmental Application Factor

A multiplicative constant used to modify a failure rate to incorporate the effects of other normal and abnormal events (IEEE Std-500, 1984).

Environmental Stress Screening

A series of tests conducted, at an early stage, under environmental stresses to disclose weak parts and workmanship defects for correction (NATO, ARMP-7, 1996).

A series of tests conducted under environmental stresses to disclose weak parts and workmanship defects for correction (USA DoD, MIL-Std–721C, 1992).

EOS

Electrical Overstress.

Equipment

A functional physical item (CENELEC, ENV50129, 1998).

An item designed and built to accomplish a specified purpose, which can be disassembled and retain its capabilities after re-assembly (ESA, ECSS-P–001A, 1997).

A generic term for an item which is designed to provide one or more services or functions to the user, or the system of which it forms a part (UK MoD, Def Stan 00–49, 1996).

Equipment Control

The equipment-specific functionality that provides the actual control capability for an equipment entity, including procedural, basic, and co-ordination control, and that is not part of the recipe (ISA, S88.01, 1995).

Equipment Entity

A collection of physical processing and control equipment and equipment control grouped together to perform a certain control function or set of control functions (ISA, S88.01, 1995).

Equipment Module

A functional group of equipment that can carry out a finite number of specific minor processing activities (ISA, S88.01, 1995).

Equipment, Off-the-Shelf

See Off-the-Shelf Equipment

Equipment, Support

See Support Equipment.

Equipment, Test

See Test Equipment.

Equipment Under Control

Equipment, machinery, apparatus or plant used for manufacturing, process, transportation, medical or other activities (IEC 61508-4, 1998).

Equivalence Class

The partition of the input domain of a program such that a test of a representative value of the class is equivalent to a test of any other value of this class (RTCA, DO178B, 1992).

Equivalence Partition

Set of input state that appear likely to be similar in that a given fault will result in a failure no matter which input state is activated (Musa et al., 1987).

Equivalent Faults

Two or more faults that result in the same failure mode (IEEE Std-610.12, 1991).

E-R Diagram

Entity-Relationship Diagram.

Ergonomics

The study of man/machine interfaces in order to minimise human errors due to mental or physical fatigue (Smith & Wood, 1989).

Error

A deviation from the intended design which could result in unintended system behaviour or failure (CENELEC, ENV50129, 1998).

A system state, resulting from a fault or human mistake, that is liable to lead to a failure (UK MoD, Def Stan 00-55, 1997; UK MoD, Def Stan 00-56, 1996).

Discrepancy between a computed, observed or measured value or condition and the true, specified or theoretically correct value or condition (ISO 2382-14, 1997; ESA, ECSS-P-001A, 1997; BSI, BS 4778-3.2, 1991; IEC 50-191, 1990; ISO 2382-2, 1976).

An error is a design flaw or deviation from a desired or intended state (Leveson, 1995).

With respect to software, a mistake in the specification, design or implementation (RTCA, DO178B, 1992).

Part of system state which is liable to lead to failure. Manifestation of a fault in a system (Laprie, 1992).

(1) The difference between a computed, observed, or measured value or condition and the true, specified, or theoretically correct value or condition. (2) An incorrect step, process, or data definition. (3) An incorrect result. (4) A human action that produces an incorrect result (IEEE Std-610.12, 1991).

Human action that results in software containing a fault (IEEE Std-982.1, 1988; IEEE Std-729, 1983).

The deviation between a calculated value (or some determined condition) and the true, specified or theoretical value (or condition) (Hölscher & Rader, 1986).

Error, Coincident

See Coincident Error.

Error Compensation

Form of error processing when erroneous state contains enough redundancy to enable correct service delivery (Laprie, 1992).

Error, Detected

See Detected Error.

Error Detecting Code

A code in which each expression conforms to specific rules of construction. If certain errors occur in an expression the resulting expression will not conform to the rules of construction and thus the presence of errors is detected (IEC 902, 1987).

Error Detection

The action of identifying that a system state is erroneous (Laprie, 1992).

Error, Dynamic

See Dynamic Error.

Error, Fatal

See Fatal Error.

Error, Human

See Human Error.

Error, Inherited

See Inherited Error.

Error, Latent

See Latent Error.

Error of Commission

The inappropriate performance of a task (or action) (BSI, BS 4778-3.1, 1991).

Error Prediction

A quantitative statement about the expected number or nature of errors in a system or component (ISO/IEC DIS 2382-7, 1996).

A quantitative statement about the expected number or nature of faults in a system or component (IEEE Std-610.12, 1991).

Error Prediction Model

In software evaluation, a model used to estimate or predict the number of remaining faults, required test time, and similar characteristics of a system (IEEE Std-610.12, 1991).

Error Processing

The actions taken in order to eliminate errors from a system (Laprie, 1992).

Error, Program

See Program Error.

Error, Programming

See Programming Error.

Error Recovery

The process of correcting or bypassing the effect of a fault or an error to allow the functional unit to continue to perform a required function (ISO 2382-14, 1997).

Form of error processing where an error-free state is substituted for an erroneous state (Laprie, 1992).

Error Recovery Software

Sections of a program, involving redundant (Parity) bits or checksums, which can recognise and correct some bit errors (Smith & Wood, 1989).

Error Seeding

Bug Seeding. Fault Seeding. The process of intentionally adding known faults to those already existing in a program for the purpose of monitoring the rate of detection and removal, and estimating the number of faults remaining in the program (ISO/IEC DIS 2382-7, 1996).

The process of intentionally adding known faults to those already in a computer program for the purpose of monitoring the rate of detection and removal, and estimating the number of faults remaining in the program (IEEE Std-610.12, 1991).

Error, Semantic

See Semantic Error.

Error, Soft

See Soft Error.

Error, Software

See Software Error.

Error, Specification

See Specification Error.

Error, Static

See Static Error.

Error, Syntactic

See Syntactic Error.

Error Tolerance

The ability of a system or component to continue normal operation despite the presence of erroneous inputs (IEEE Std-610.12, 1991).

Error, Transient

See Transient Error.

ESD

(1) Emergency Shutdown. (2) Electrostatic Discharge.

ESS

See Environmental Stress Screening.

Estimation, Parameter

See Parameter Estimation.

Estimation, Risk

See Risk Estimation.

ETA

Event Tree Analysis.

EUC

See Equipment Under Control.

EUT

Equipment Under Test.

Evaluating

Determination of characteristic values or relationships of a process from measured values of process variables (IEC 50-351, 1998).

Evaluation

A systematic examination of the extent to which an entity is capable of fulfilling specified requirements (UK MoD, Def Stan 05-95, 1995).

The process of defining whether an item or activity meets specified criteria (USA DoD, MIL-Std-498, 1994; USA DoD, MIL-Std-2167A, 1988).

The assessment of an IT system or product against defined evaluation criteria (UK DTI, ITSEC, 1991).

Determination of fitness for use (IEEE Std-1074, 1991).

Evaluation, Risk

See Risk Evaluation.

Evaluation, Security

See Security Evaluation.

Event

The actual or contemplated occurrence of something (ESA, ECSS-P-001A, 1997).

A significant happening that may originate in the system or its operating environment (UK MoD, Def Stan 00-58, 1996).

A significant happening that may originate in the system or its domain (UK MoD, Def Stan 00-56, 1996).

Event, Downing

See Downing Event.

Event, Initiating

See Initiating Event.

Event Maintenance

One or more maintenance actions required to effect corrective and preventive maintenance due to any type of failure or malfunction, false alarm or scheduled maintenance plan (USA DoD, MIL-Std-721C, 1992).

Event, Systematic

See Systematic Event.

Event, Top

See Top Event.

Event Tree Analysis

A method for illustrating the intermediate and final outcomes which may arise after the occurrence of a selected initial event (Jones, 1992).

Examination

An element of inspection consisting of investigation, without the use of special laboratory appliances, or procedures, supplies or services to determine conformance to those specified requirements which can be determined by such investigations (USA DoD, MIL-Std-109C, 1994).

Exception

A condition that may arise during execution of a program, that may cause a deviation from the normal execution sequence, and for which means exist to define, raise, recognise, ignore, or handle it (ISO/IEC DIS 2382-7, 1996).

An exception is an internal software event that can occur to indicate that the software is in error. The program may look out for exceptions and take remedial action if they occur; this is termed handling the exception. A typical exception might be one to indicate that arithmetic overflow has occurred (SSP 84, 1996).

An event that causes suspension of normal program execution (IEEE Std-610.12, 1991).

Exception, Data

See Data Exception.

Exception Handling

Those functions that deal with plant or process contingencies and other events which occur outside the normal or desired behaviour of batch control (ISA, S88.01, 1995).

Excess Strength

Method to achieve reliability. Deliberate design to withstand a higher stress than that which is required to be endured. Small increases in strength for a given applied stress result in substantial decreases in failure rate. This applies equally to mechanical and electrical designs (Smith, 1981).

Executable Prototype

An executable form of a formal specification that is used to validate the software specification (UK MoD, Def Stan 00-55, 1997).

Execution Profile

A representation of the absolute or relative execution frequencies or execution times (2) of the instructions or of the statements of a program (ISO/IEC DIS 2382-7, 1996).

Execution, Symbolic

See Symbolic Execution.

Execution Time

Run time. (1) Any instant at which the execution of a particular program takes place. (2) The amount of time needed for the execution of a particular program (ISO/IEC DIS 2382-7, 1996).

The amount of elapsed time or processor time used in executing a computer program (IEEE Std-610.12, 1991).

The time that is actually spent by a processor in executing the instructions of that program (Musa et al., 1987).

External Fault

Fault resulting from environmental interference or interaction (Laprie, 1992).

F

Fail-Safe

A fail-safe system cannot allow a more permissive condition than that existing in the absence of a failure (CENELEC, prEN50126, 1998).

The capability to go to a predetermined safe state in the event of a specific malfunction (ISA, S84.01, 1996).

A design feature that ensures that the system remains safe or in the event of a failure will cause the system to revert to a state which will not cause a mishap (USA DoD, MIL-Std-882C, 1993).

A concept that defines the failure direction of a component/system as a result of specific malfunctions. The failure direction is toward a safer or less hazardous condition (CCPS, 1993).

Pertaining to a system or component that automatically places itself in a safe operating mode in the event of a failure (IEEE Std-610.12, 1991).

A design property of an item which prevents its failures from resulting in critical faults (BSI, BS 4778-3.2, 1991; IEC 50-191, 1990).

The capacity of the system to remain in a safe condition when a failure occurs or to skip directly into another safe condition (VDE 0801, 1990).

Designing so that failure of power, control circuits, structural members or other components will not endanger the process under control or people operating the system or other people in the vicinity (IEC 902, 1987).

Fail-Safe Design

The design of a guard, elevator, equipment, hoist, etc. such that, should it fail or become inoperative, it will do so in a safe position or condition; or the design of a hazard control such that its failure will not result in additional exposure to the hazard (ASSE, 1988).

Fail-Safe Fault

A fault which results in no deterioration of safety (Jones, 1992).

Fail-Safe Shutdown

The ability of a Programmable Controller System to have its outputs assume a predefined state within a specified delay after detecting the occurrence of a power supply voltage drop or an internal failure (IEC 1131-1, 1992).

Fail-Safe State

A state of the system that cannot result in an accident, although some other important goals of the system, such as availability, may be compromised (UK MoD, Def Stan 00-56, 1996).

Fail-Safe System

System whose failures can only be, or are to an acceptable extent, benign, failures (Laprie, 1992).

Fail-Silent System

System whose failures can only be, or are to an acceptable extent, crashing failures (Laprie, 1992).

Fail-Soft

Pertaining to a functional unit continuing to function in a degraded mode despite faults or manual out-of-limits operations (ISO 2382–14, 1997).

Pertaining to a system or component that continues to provide partial operational capability in the event of certain failures (IEEE Std–610.12, 1991).

Fail-Stop System

System whose failures can only be, or are to an acceptable extent, stopping failures (Laprie, 1992).

Fail-to-Danger

An equipment fault which inhibits or delays actions to achieve a safe operational state should a demand occur. The fail-to-danger fault has a direct and detrimental effect on safety (CCPS, 1993).

Failure

A deviation from the specified performance of a system. A failure is the consequence of a fault or error in the system (CENELEC, ENV50129, 1998).

Termination of the ability of a functional unit to perform a required function (IEC 61508–4, 1998; ISO 2382–14, 1997).

The inability of a system or component to fulfil its operational requirements. Failures may be systematic or due to physical change (UK MoD, Def Stan 00–55, 1997; UK MoD, Def Stan 00–58, 1996; UK MoD, Def Stan 00–56, 1996).

The termination of the ability of an item to perform a required function (ESA, ECSS-P-001A, 1997; BSI, BS 4778–3.2, 1991; IEC 50–191, 1990; O'Connor, 1981).

The inability of an item to perform within previously specified limits (NATO, ARMP-7, 1996; USA DoD, MIL-Std–109C, 1994).

Failure is the non-performance or inability of the system or component to perform its intended function for a specified time under specified environmental conditions (Leveson, 1995).

The delivered service deviates from the specified service (Korteweg, 1994).

The event, or inoperable state, in which any item or part of an item does not, or would not, perform as previously specified (USA DoD, MIL-Std–721C, 1992).

The termination of the ability of an item or equipment to perform its required function (OREDA, 1992; IEEE Std–500, 1984).

The inability of a system or system component to perform a required function within specified limits. A failure may be produced when a fault is encountered (RTCA, DO178B, 1992).

Deviation of the delivered service from compliance with the specification. Transition from correct service delivery to incorrect service delivery (Laprie, 1992).

The inability of a system or component to perform its required functions within specified performance requirements (IEEE Std–610.12, 1991).

The termination of the execution of an established task from a unit as a result of a cause which is located in the unit itself and within the framework of the permitted working conditions (VDE 0801, 1990).

(1) The termination of the ability of a functional unit to perform its required function. (2) An event in which a system or system component does not perform a required function within specified limits. A failure may be produced when a fault is encountered (IEEE Std–982.1, 1988; IEEE Std–729, 1983).

The termination of the ability of a functional unit to perform its required function (IEC 902, 1987).

Behaviour of a unit which is carrying out some task which is not in accord with its intended or specified function (Hölscher & Rader, 1986).

Termination of the ability of an item to perform its specified function. Or, non-conformance to some defined performance criteria (Smith, 1981).

Failure, Active

See Active Failure.

Failure Analysis

The logical, systematic examination of a failed item to identify and analyse the failure mechanism, the failure cause and the consequences of failure (NATO, ARMP-7, 1996; IEC 50-191, 1990).

The investigation into the degree of imperfection to which an item has degenerated (or failed) when measured against a previously specified limit (USA DoD, MIL-Std-109C, 1994).

Failure, Arbitrary

See Failure (Laprie, 1992).

Failure, Benign

See Benign Failure.

Failure, Catastrophic

See Catastrophic Failure.

Failure Cause

The circumstances during design, manufacture or use which have lead to a failure (CENELEC, prEN50126, 1998; BSI, BS 4778-3.2, 1991; IEC 50-191, 1990).

The identified original cause of the failure; the circumstances during design, manufacturing, assembly, installation or use that have led to failure (OREDA, 1992).

Failure, Common Cause

See Common Cause Failure.

Failure, Common Mode

See Common Mode Failure.

Failure, Consistent

See Consistent Failure.

Failure, Crash

See Crash Failure.

Failure, Critical

See Critical Failure.

Failure, Dangerous

See Dangerous Failure.

Failure, Degraded

See Degraded Failure.

Failure, Dependent

See Dependent Failure.

Failure Diagnosis

The investigation of the facts available (such as an investigation of the failed item itself, of failure mode(s), and of contributory causes) to determine the nature of the failure (USA DoD, MIL-Std-109C, 1994).

Failure, Early Life

See Early Life Failure.

Failure Effect

The consequence(s) a failure mode has on the operation, function, or status of an item. Failure efforts are classified as local effect, next higher-level, and end effect (USA DoD, MIL-Std-721C, 1992).

Failure, Hard

See Hard Failure.

Failure, Incipient

See Incipient Failure.

Failure, Inconsistent

See Inconsistent Failure.

Failure, Independent

See Independent Failure.

Failure Integrity, Systematic

See Systematic Failure Integrity.

Failure Intensity

Failures per unit time, the derivative with respect to time of the mean value function of failures (Musa et al., 1987).

Failure, Intermittent

See Intermittent Failure.

Failure, Internal

See Internal Failure.

Failure Interval

Time between failures (Musa et al., 1987).

Failure, Mean Time Between

See Mean Time Between Failure.

Failure, Mean Time To

See Mean Time To Failure.

Failure Mechanism

The physical, chemical, electrical, thermal or other process which results in failure (USA DoD, MIL-Std-721C, 1992).

The physical, chemical, or other process which results in failure. Note: the circumstance that induces or activates the process is termed the root cause of the failure (IEEE Std-500, 1984).

Failure Mode

The predicted or observed results of a failure cause on a stated item in relation to the operating conditions at the time of the failure (CENELEC, prEN50126, 1998).

The observable effect of the mechanism through which the failure occurs, e.g. short-circuit, open-circuit, fracture, excessive wear (ESA, ECSS-P-001A, 1997).

The consequence of the mechanism through which the failure occurs, i.e., short, open, fracture, excessive wear (NATO, ARMP-7, 1996; USA DoD, MIL-Std-721C, 1992).

The manner in which a component fails (Jones, 1992).

The effect by which a failure is observed on the failed item. The failure modes describe the variations of the item state that result from failures (OREDA, 1992).

The effect by which a failure is observed (BSI, BS 4778–3.1, 1991).

The physical or functional manifestation of a failure (IEEE Std–610.12, 1991).

The effect by which a failure is observed in a system component (IEC 812, 1985).

The effect by which a failure is observed to occur (IEEE Std–500, 1984).

The outward appearance of a failure (Smith, 1981).

Failure Mode and Effects Analysis

A procedure by which each potential failure mode of a component, equipment or subsystem in a system is analysed to determine the results or effects thereof on the overall system and to classify each potential failure mode according to its severity (NATO, ARMP–7, 1996).

A procedure by which each potential failure mode in a system is analysed to determine the results or effects thereof on the system and to classify each potential failure mode according to its severity (USA DoD, MIL-Std–721C, 1992).

A process for hazard identification where all known failure modes of components or features of a system are considered in turn and undesired outcomes are noted (Jones, 1992).

Failure Mode, Common

See Common Failure Mode.

Failure Mode Types

(1) Catastrophic. A failure which is both sudden and complete. (2) Degraded. A failure which is gradual, partial, or both. (3) Incipient. An

imperfection in the state or condition of an item or equipment so that a degraded or catastrophic failure can be expected to result if corrective action is not taken (IEEE Std–500, 1984).

Failure, Non-Chargeable

See Non-Chargeable Failure.

Failure, Non-Relevant

See Non-Relevant Failure.

Failure, Omission

See Omission Failure.

Failure, Premature

See Early Life Failure (NATO, ARMP–7, 1996).

Failure, Primary

See Primary Failure.

Failure, Random

See Random Failure.

Failure Rate

The limit, if this exists, of the ratio of the conditional probability that the instant of time, T, of a failure of a product falls within a given time interval (t, t+δt) and the length of this interval δt, when δt tends towards zero, given that the item is in an up-state at the start of the time interval (CENELEC, prEN50126, 1998).

The number failures of an item per unit measure of life, expressed in hours, cycles, kilometres, events as applicable to the item (NATO, ARMP–7, 1996).

The total number of failures within an item population, divided by the total number of life units expended by that population, during a particular measurement interval under stated conditions (USA DoD, MIL-Std-721C, 1992).

The rate at which failures occur as a function of time or as function of demands (OREDA, 1992).

The ratio of the number of failures of a given category to a given unit of measure (IEEE Std-610.12, 1991).

The expected number of failures of a given type, per item, in a given time interval (for example, valve failures per million valve hours). Note: For cyclic items or equipment insert 'in a given number of operating cycles' (IEEE Std-500, 1984).

The number of failures of an item per unit time (per hour, cycle, operation etc.). This can be applied to (1) Observed failure rate: as computed from a sample. (2) Assessed failure rate: as inferred from sample information. (3) Extrapolated failure rate: projected to other stress levels (Smith, 1981).

Failure Rate, Observed

See Observed Failure Rate.

Failure Ratio

The probability that an item will fail or that a trial will unsuccessful under stated conditions (IEC 1123, 1991).

Failure, Revealed

See Revealed Failure.

Failure Robust, Revealed

See Revealed Failure Robust.

Failure Robust, Unrevealed

See Unrevealed Failure Robust.

Failure, Safe

See Safe Failure.

Failure, Secondary

See Secondary Failure.

Failure Severity

Grade of the failure consequences upon the system environment (Laprie, 1992).

Classification of a failure by its operational impact (Musa et al., 1987).

Failure, Software

See Software Failure.

Failure, Stopping

See Stopping Failure.

Failure, Sudden

See Sudden Failure.

Failure, Systematic

See Systematic Failure.

Failure Time

Accumulated elapsed time at which a failure occurs (Musa et al., 1987).

Failure, Timing

See Timing Failure.

Failure, Unrevealed

See Unrevealed Failure.

Failure, Value

See Value Failure.

Failure, Wearout

See Wearout Failure.

Failures, Multiple

See Multiple Failures.

Failures, Pattern

See Pattern Failures.

Failures, Sequential

See Sequential Failures.

Failures, Simultaneous

See Simultaneous Failures.

Failures, Total

See Total Failures.

False Shutdown

See Spurious Trip (ISA, S84.01, 1996).

FAR

Fatal Accidents Rate.

FAT

Factory Acceptance Test.

Fatal Accidents Rate

The number of deaths that have occurred or are predicted to occur in a defined group, in a given environment, during 100 million hours of total exposure (Jones, 1992).

Fatal Error

An error that results in the complete inability of a system or component to function (IEEE Std-610.12, 1991).

Fault

An abnormal condition that could lead to an error in a system. A fault can be random or systematic (CENELEC, ENV50129, 1998).

Abnormal condition that may cause a reduction in, or loss of, the capability of a functional unit to perform a required function (IEC 61508-4, 1998; ISO 2382-14, 1997).

Immediate cause of failure (e.g. maladjustment, misalignment, defect, etc.) (USA DoD, MIL-Std-721C, 1992).

A manifestation of an error in software, a mistake in specification, design or implementation (RTCA, DO178B, 1992).

The state of an item characterised by inability to perform a required function, excluding the inability during preventive maintenance or other planned actions, or due to lack of external resources (BSI, BS 4778-3.2, 1991; IEC 50-191, 1990).

(1) A defect in a hardware device or component. (2) An incorrect step, process, or data definition in a computer program (IEEE Std–610.12, 1991).

(1) An accidental condition that causes a functional unit to fail to perform its required function. (2) A manifestation of an error in software. A fault, if encountered, may cause a failure. Synonymous with bug (IEEE Std–982.1, 1988; IEEE Std–729, 1983).

An accidental condition that causes a functional unit to fail to perform its required function (IEC 902, 1987).

Defective, missing, or extra instruction or set of related instructions that is the cause of one or more actual or potential failure types (Musa et al., 1987).

Inadmissible characteristic of a given system that can (but need not) lead to the failure of that system (Hölscher & Rader, 1986).

Fault, Accidental

See Accidental Fault.

Fault, Active

See Active Fault.

Fault Avoidance

Use of techniques and procedures which aim to avoid the introduction of faults during any phase of the safety life cycle of the safety-related system (IEC 61508–4, 1998).

Methods and techniques aimed at producing a fault-free system. Fault prevention and fault removal (Laprie, 1992).

Fault-Based Testing

Testing aimed at revealing specific classes of faults (Laprie, 1992).

Fault, Common Cause

See Common Cause Fault.

Fault, Common Mode

See Common Mode Fault.

Fault Coverage

The proportion of faults of an item that can be recognised under given conditions (BSI, BS 4778-3.2, 1991; IEC 50-191, 1990).

Fault, Covert

See Covert Fault.

Fault, Critical

See Critical Fault.

Fault, Data-Sensitive

See Data-Sensitive Fault.

Fault, Degradation

See Degradation Fault.

Fault, Design

See Design Fault.

Fault Diagnosis

The action of determining the cause of an error in location and nature (Laprie, 1992).

Fault Exposure Ratio

Proportionality factor that relates failure intensity with the rate at which faults would be encountered if the program were executed linearly (Musa et al., 1987).

Fault, External

See External Fault.

Fault, Fail-Safe

See Fail-Safe Fault.

Fault-Finding Testing

Testing whose purpose is finding faults (Laprie, 1992).

Fault Forecasting

Methods and techniques aimed at estimating the present number, the future incidence, and the consequence of failures (Laprie, 1992).

Fault, Hard

See Hard Fault.

Fault, Independent

See Independent Fault.

Fault, Inherent

See Inherent Fault.

Fault, Intentional

See Intentional Fault.

Fault, Intermittent

See Intermittent Fault.

Fault, Internal

See Internal Fault.

Fault Isolation

The process of determining the location of a fault to the extent necessary to effect repair (USA DoD, MIL-Std-721C, 1992).

Fault, Latent

See Latent Fault.

Fault Localisation

The process of determining the approximate location of a fault (USA DoD, MIL-Std-721C, 1992).

Actions taken to identify the faulty subitem or subitems at the appropriate indenture level (BSI, BS 4778-3.2, 1991; IEC 50-191, 1990).

Fault Masking

The result of applying error compensation systematically, even in the absence of error (Laprie, 1992).

The condition in which a fault exists in a subitem of an item but cannot be recognised because of a feature of the item or because of another fault of that subitem or of another sub-item (BSI, BS 4778-3.2, 1991; IEC 50-191, 1990).

A condition in which one fault prevents the detection of another (IEEE Std-610.12, 1991).

Fault Mode

One of the possible states of a faulty product for a given required function (CENELEC, prEN50126, 1998).

Fault, Operational

See Operational Fault.

Fault, Overt

See Overt Fault.

Fault Passivation

The actions taken in order that a fault cannot be activated (Laprie, 1992).

Fault, Passive

See Passive Fault.

Fault, Permanent

See Permanent Fault.

Fault, Persistent

See Persistent Fault.

Fault, Physical

See Physical Fault.

Fault Prevention

Methods and techniques aimed at preventing fault occurrence or introduction (Laprie, 1992).

Fault, Random

See Random Fault.

Fault Reaction Period

The length of time from the discovery of a fault to the achievement of a safe state (Hölscher & Rader, 1986).

Fault Reduction Factor

Net reduction in faults per failure experienced (Musa et al., 1987).

Fault Removal

Methods and techniques aimed at reducing the presence (number, seriousness) of faults (Laprie, 1992).

Fault, Revealed

See Revealed Fault.

Fault Seeding

See Error Seeding (ISO/IEC DIS 2382-7, 1996; IEEE Std-610.12, 1991).

Fault, Soft

See Soft Fault.

Fault, Solid

See Solid Fault.

Fault, Systematic

See Systematic Fault.

Fault, Temporary

See Temporary Fault.

Fault Tolerance

Ability of a functional unit to continue to perform a required function in the presence of faults or errors (IEC 61508–4, 1998; ISO 2382–14, 1997).

The attribute of an item that makes it able to perform a required function in the presence of certain given subitem faults (ESA, ECSS-P–001A, 1997; NATO, ARMP–7, 1996; BSI, BS 4778–3.2, 1991; IEC 50–191, 1990).

Built-in capability of a system to provide continued correct execution of its assigned function in the presence of a limited number of hardware and software faults (ISA, S84.01, 1996).

That property of a system which permits it to carry out its assigned function in the presence of one or more faults in the hardware or software (CCPS, 1993).

Methods and techniques aimed at providing a service complying with the specification in spite of faults (Laprie, 1992).

(1) The ability of a system or component to continue normal operation despite the presence of hardware or software faults. (2) The number of faults a system or component can withstand before normal operation is impaired. (3) Pertaining to the study of errors, faults, and failures, and of methods for enabling systems to continue normal operation in the presence of faults (IEEE Std--610.12, 1991).

Incorporating design features to protect against effects of defects (Musa et al., 1987).

Fault Tolerance Period

The length of time during which a process can be affected by incorrect control signals without moving into a dangerous state (Hölscher & Rader, 1986).

Fault, Transient

See Transient Fault.

Fault Treatment

The actions taken in order to prevent a fault from being re-activated (Laprie, 1992).

Fault Tree Analysis

An analysis to determine which fault modes of the product, subproducts or external events, or combinations thereof, may result in a stated fault mode of the product, presented in the form of a fault tree (CENELEC, prEN50126, 1998).

An FTA provides a diagrammatic means of showing the logical relationship between a particular system failure mode and the basic failure modes (NATO, ARMP--7, 1996).

A method for representing the logical combinations of various system states which lead to a particular outcome (Jones, 1992).

FTA is a top-down or deductive method of determining the combination of failures and operating circumstances which could cause a hazard. Basic events, those for which sufficient information already exists, e.g. a frequency, failure rate or probability of failure is known, are combined by the laws of Boolean algebra to identify which basic events cause a failure mode of a PES (HSE, 1991).

Fault, Unrevealed

See Unrevealed Fault.

Faults, Equivalent

See Equivalent Faults.

Faults, Related

See Related Faults.

FCA

Functional Configuration Audit.

FDT

See Fractional Dead Time.

Feature, Safety

See Safety Feature.

Feedback Control

A method by which one or more controlled variables (pressures, temperature, current, speed, power, etc.) are made to obey a command signal, whether constant or varying, according to a prescribed law, as a result of measurement of the variable(s) in question (IEC 1131-1, 1992).

Control in which the control action is made to depend persistently on the measurement of the controlled variable (IEC 902, 1987).

FHA

Functional Hazard Assessment.

Field Devices

Equipment connected to the field side of the Safety Instrumented System I/O terminals. Such equipment includes field wiring, sensors, final control elements and those operator interface devices hard-wired to Safety Instrumented System I/O terminals (ISA, S84.01, 1996).

Field Test

A compliance test or determination test made in the field where operating, environmental, maintenance and measurement conditions present at the time of the test are recorded (BSI, BS 4778-3.2, 1991; IEC 50-191, 1990).

Final Control Element

A device that manipulates a process variable to achieve control (CCPS, 1993).

Element of the forward path which directly changes the manipulated value (IEC 902, 1987).

Firmware

The combination of a hardware device and computer instructions and data that reside as read-only software on that device (UK MoD, Def Stan 00-55, 1997).

Hardware that contains a computer program or data that cannot be changed in its user environment. The computer program and data contained in firmware are classified as software; the circuitry containing the computer program and data is classified as hardware (ESA, ECSS-P-001A, 1997).

Special purpose memory units containing software embedded in protected memory required for the operation of programmable electronics (ISA, S84.01, 1996).

An ordered set of instructions and associated data stored in a way that is functionally independent of main storage, usually in a ROM (ISO 2382-1, 1993).

The combination of a hardware device and computer instructions and data that reside as read-only software on that device (IEEE Std-610.12, 1991).

Mostly basic functions which are implemented by the computer manufacturer with procedures from software technology in order to realise a machine language or other machine functions. Firmware cannot be modified by the user by giving software commands. As far as the user is concerned, the firmware has the same function as hardware (VDE 0801, 1990).

The combination of a hardware device and computer instructions or computer data that reside as read-only software on the hardware device. The software cannot be readily modified under program control (USA DoD, MIL-Hdbk-347, 1990; USA DoD, MIL-Std-2167A, 1988).

FIT

Failures In Time. Failure unit: failures per 10^9 hours.

Fitness for Purpose

Ability of a product, process or service to serve a defined purpose under specific conditions (ISO/IEC, Guide 2, 1996).

Fitness for Use

The effectiveness of the design, manufacturing, and support processes in delivering a system that meets the operational requirements under all anticipated operational conditions (USA DoD, MIL-Std-109C, 1994).

FMEA

See Failure Mode and Effects Analysis.

FMECA

Failure Mode, Effects, and Criticality Analysis.

Forcing

A PES engineering station function that provides the capability to override the application program and to change the states of inputs and outputs (ISA, S84.01, 1996).

Forecasting, Fault

See Fault Forecasting.

Formal

A description applied to a specification or process expressed in a rigorous mathematical form (SSCP 85, 1995).

Formal Argument

Formal proofs or rigorous arguments (UK MoD, Def Stan 00-55, 1997).

Formal Design

The part of the software design written using a formal notation (UK MoD, Def Stan 00-55, 1997).

Formal Design Review

A review, at specific development stages, of the technical aspects of the software components, including documentation, to ensure compliance with requirements, functional specifications, standards, and other requirements appropriate to that stage (CSA, CAN/CSA-Q396.1.1, 1989).

Formal Language

A language whose rules are explicitly established prior to its use (IEEE Std-610.12, 1991).

Formal Method

A software specification and production method, based on mathematics, that comprises: a collection of mathematical notations addressing the specification, design and development processes of software production; a well-founded logical inference system in which formal verification proofs and proofs of other properties can be formulated; and a methodological framework within which software may be developed from the specification in a formally verifiable manner (UK MoD, Def Stan 00-55, 1997).

Descriptive notations and analytical methods used to construct, develop and reason about mathematical models of system behaviour (RTCA, DO178B, 1992).

Formal Notation

The mathematical notation of a formal method (UK MoD, Def Stan 00–55, 1997).

Formal Proof

The discharge of a proof obligation by the construction of a complete mathematical proof (UK MoD, Def Stan 00–55, 1997).

Formal Qualification Review

The test, inspection, or analytical process by which a group of configuration items comprising a system is verified to have met specific contractual performance requirements (IEEE Std–610.12, 1991).

Formal Specification

The part of the software specification written using a formal notation (UK MoD, Def Stan 00–55, 1997).

A specification written in a formal notation, often for use in correctness proving (ISO/IEC DIS 2382–7, 1996).

(1) A specification written and approved in accordance with established standards. (2) A specification written in a formal notation, often for use in proof of correctness (IEEE Std–610.12, 1991).

Formal Testing

Testing conducted in accordance with test plans and procedures that have been reviewed and approved by a customer, user, or designated level of management (IEEE Std–610.12, 1991).

Forward Recovery

Form of error recovery where the erroneous state transformation consists of finding a new state (Laprie, 1992).

F-N Curve

A plot showing, for a specified hazard, the frequency of all events causing a stated degree of harm to N or more people, against N (Jones, 1992).

FQR

See Formal Qualification Review.

FRACAS

Failure Reporting, Analysis, and Corrective Action System.

Fractional Dead Time

The mean fraction of time in which a component or system is unable to operate on demand (Jones, 1992).

Free Time

The portion of up-time or down-time during which the operational use of the system is not required (Martz & Waller, 1982).

FTA

See Fault Tree Analysis.

Function

An aspect of the intended behaviour of the system (UK MoD, Def Stan 00–58, 1996; UK MoD, Def Stan 00–56, 1996).

What a system is intended for (Laprie, 1992).

The action or actions which an item is designed to perform (USA DoD, MIL-Std-973, 1992).

A specific purpose of an entity, or its characteristic action (IEEE Std-1074, 1991; IEEE Std-1063, 1988).

(1) A defined objective or characteristic action of a system or component. (2) A software module that performs a specific action, is invoked by the appearance of its name in an expression, may receive input values, and returns a single value (IEEE Std-610.12, 1991).

Function Block Diagram Language

A programming language using function block diagrams for representing the application program for a Programmable Controller System (IEC 1131-1, 1992).

Function Chart, Sequential

See Sequential Function Chart.

Function, Real-Time

See Real-Time Function.

Function, Safety

See Safety Function.

Functional Baseline

The configuration documentation formally designated at the beginning of project definition prescribing: (1) all necessary functional characteristics; (2) test requirements; (3) the necessary interface characteristics with associated configuration items; (4) key lower-level configuration items, if any; and (5) design constraints (UK MoD, Def Stan 05-57, 1993).

The initially approved documentation describing a system's or item's functional characteristics and the verification required to demonstrate the achievement of those specified functional characteristics (USA DoD, MIL-Hdbk–347, 1990; USA DoD, MIL-Std–480B, 1988).

Functional Block

Delimited part of a system, comprising at least one function, characterised by its having a minimal number of well-defined input and output signals (IEC 902, 1987).

Functional Design

(1) The process of defining the working relationships among the components of a system. (2) The result of the process in (1) (IEEE Std–610.12, 1991).

Functional Hazard

A systematic, comprehensive examination of system functions to identify and classify potentially hazardous failure conditions and to describe them in functional and operational terms.

Functional Requirement

The statements that define the essential features of the software components, along with the technical constraints and conditions that are to be met (CSA, CAN/CSA-Q396.1.1, 1989).

Functional Safety

Part of the overall safety relating to the EUC and the EUC control system which depends on the correct functioning of the E/E/PE safety-related systems, other technology safety-related systems and external risk reduction facilities (IEC 61508–4, 1998).

Functional Safety Assessment

Investigation, based on evidence, to judge the functional safety achieved by one or more E/E/PE safety-related systems, other technology safety-related systems or external risk reduction facilities (IEC 61508–4, 1998).

Functional Safety Audit

Systematic and independent examination to determine whether the procedures specific to the functional safety requirements comply with the planned arrangements are implemented effectively and are suitable to achieve the specified objectives (IEC 61508-4, 1998).

Functional Test

An empirical test routine designed to exercise an item such that all aspects of the software are brought into use (Smith, 1981).

Functional Testing

Periodic activity to verify that the Safety Instrumented System is operating per the Safety Requirement Specifications Testing (ISA, S84.01, 1996).

A systematic sequence of steps used to check the condition of a completed end item to test its fitness for use (USA DoD, MIL-Std-109C, 1994).

Form of testing where the testing inputs are selected according to criteria relating to the system's function (Laprie, 1992).

(1) Testing that ignores the internal mechanism of a system or component and focuses solely on the outputs generated in response to selected inputs and execution conditions. (2) Testing conducted to evaluate the compliance of a system or component with specified functional requirements (IEEE Std-610.12, 1991).

Functional Unit

An entity of hardware or software, or both, capable of accomplishing a specified purpose (IEC 61508-4, 1998; ISO 2382-14, 1997; ISO 2382-1, 1993).

A functionally distinct part of a basic component (UK DTI, ITSEC, 1991).

G

Generator, Test Case

See Test Case Generator.

Generic Software

Generic software is software that can be tailored to process data of various types. A sort routine would typically be generic because it should sort items of any type provided there is a suitable order defined for them (SSP 84, 1996).

Glass Box

(1) A system or component whose internal contents or implementation are known. (2) Pertaining to an approach that treats a system or component as in (1) (IEEE Std-610.12, 1991).

Glass-Box Testing

See Structural Testing (IEEE Std-610.12, 1991).

Graceful Degradation

The sufficiently long interval between the occurrence of a potential failure, which produces degraded capability, and its decay into functional failure to allow either: (a) maintenance/repair intervention to prevent functional failure; or (b) action to be taken to avoid or reduce the consequences of functional failure (UK MoD, Def Stan 00-49, 1996).

A design feature whereby a system continues to operate, albeit at a reduced efficiency or with fewer functions available, in the presence of failures (Smith & Wood, 1989).

Graphical Language

A programming language based upon graphical representation (IEC 1131-1, 1992).

Graphical User Interface

Graphical User Interface. An interface to an application program which is intended to be easy to use and typically uses a mouse to point and click at various symbols (icons) on the screen in order to give instructions to the program (SSP 84, 1996).

GUI

See Graphical User Interface.

Guidewords

A list of words applied to system items or functions in a hazard study to identify undesired deviations (Jones, 1992).

H

Hard Failure

A failure that results in complete shutdown of a system (IEEE Std-610.12, 1991).

Hard Fault

Fault necessitating passivation (Laprie, 1992).

Hardware

End items, physical equipment, or reparable items (USA DoD, MIL-Std-109C, 1994).

All or part of the physical components of an information processing system (ISO 2382-1, 1993).

Physical equipment used in data processing as opposed to computer program, procedures, rules and associated documentation (IEC 1131-1, 1992).

Physical equipment used to process, store, or transmit computer programs or data (IEEE Std–610.12, 1991).

Complete or part of the (material) equipment from the computing systems (VDE 0801, 1990).

Physical material directly involved in performing a specific function (IEC 902, 1987).

The totality or part of the fittings constituting the system (Hölscher & Rader, 1986).

Hardware, Application

See Application Hardware.

Hardware, Computer

See Computer Hardware.

Hardware Configuration

See Architecture (ISA, S84.01, 1996).

Hardware Failure, Random

See Random Hardware Failure.

Hardware/Software Integration

The process of combining the software into the target computer (RTCA, DO178B, 1992).

Hard-Wire

A dedicated discrete electrical circuit (USA DoD, MIL-Hdbk–272A, 1993).

Hard-Wired

Electrical connections accomplished without the use of software or firmware (ISA, S84.01, 1996).

Harm

Physical injury or damage to the health of people either directly, or indirectly as a result of damage to property or to the environment (IEC 61508-4, 1998).

Physical injury or damage to health, property or the environment (IEC 300-3-4, 1996).

Harness, Test

See Test Harness.

HAST

Highly Accelerated Stress Test.

HAT

Hardware Acceptance Test.

Hazard

Potential source of harm (IEC 61508-4, 1998).

A condition that could lead to an accident (CENELEC, ENV50129, 1998).

A physical situation with a potential for human injury (CENELEC, prEN50126, 1998).

A condition, associated with the design, operation or environment of a system, that has the potential for harmful consequences (ESA, ECSS-P-001A, 1997).

A physical situation, often following from some initiating event, that can lead to an accident (UK MoD, Def Stan 00–58, 1996; UK MoD, Def Stan 00–56, 1996).

A chemical or physical condition that has the potential for causing damage to people, property, or the environment (ISA, S84.01, 1996; CCPS, 1993).

A state or set of conditions of a system (or an object) that, together with other conditions in the environment of the system (or object), will lead inevitably to an accident (loss event) (Leveson, 1995).

Source of potential harm or a situation with a potential for harm (IEC 300–3–9, 1995).

A physical situation with a potential for human injury/equipment and production loss/environmental pollution (Korteweg, 1994).

A condition that is prerequisite to a mishap (USA DoD, MIL-Std–882C, 1993).

A situation where a person may be injured (HSE, 1991).

Hazard Analysis

The process of identifying the hazards which a product or its use can cause (CENELEC, ENV50129, 1998).

An iterative process composed of identification and evaluation of hazards associated with the computer system, to enable them to be eliminated or, if that is not practical, to assist in the reduction of the associated risks to an acceptable level (UK MoD, Def Stan 00–56, 1996).

An analysis for the purpose of exploring the hazards which may be caused by the system or which may affect the system (UK MoD, Def Stan 00–58, 1996).

Determination of possible failure modes of a system and the consequences of such failures, particularly those that could risk human life (SSCP 85, 1995).

The identification of undesired events that lead to the materialisation of a hazard, the analysis of the mechanisms by which these undesired events could occur and usually the estimation of the extent, magnitude and likelihood of any harmful effects (Jones, 1992).

Hazard and Operability Study

A formal systematic examination, by a team under the management of a trained leader, of the design intentions of a new or existing system or parts of a system, to identify hazards, maloperation or malfunction of individual entities within the system and the consequences on the system as a whole and on its environment. It typically includes several HAZOP study meetings (UK MoD, Def Stan 00-58, 1996).

A study carried out by application of guidewords to identify all deviations from design intent with undesirable effects for safety or operability (Jones, 1992).

Hazard, Functional

See Functional Hazard.

Hazard Log

The document in which all safety management activities, hazards identified, decisions made and solutions adopted are recorded or referenced (CENELEC, ENV50129, 1998; CENELEC, prEN50126, 1998).

The document, established in accordance with Def Stan 00-56, that acts as a record and directory for the safety justification for the system (UK MoD, Def Stan 00-55, 1997).

Hazard Rate

The frequency at which hazards occur. Dimension (time^-1). Hazard Rate = Demand Rate × Probability of Failure on Demand (Korteweg, 1994).

Probability density (per unit time) of failure given that failure has not occurred up to the present (Musa et al., 1987).

The instantaneous failure rate of an item or equipment or its conditional probability of failure versus time (IEEE Std–500, 1984).

Hazard Rate, Item or Component

The instantaneous failure rate of an item or equipment or its conditional probability of failure versus time (IEEE Std–500, 1984).

Hazard Severity

An assessment of the consequences of the worst credible mishap that could be caused by a specific hazard (USA DoD, MIL-Std–882C, 1993).

Hazard, Software

See Software Hazard.

Hazard Survey

The total effort involved in an assessment of the hazards from an installation and their means of control (Jones, 1992).

Hazard, System

See System Hazard.

Hazardous Event

Hazardous situation which results in harm (IEC 61508–4, 1998).

Event which can cause harm (IEC 300–3–9, 1995).

Hazardous Situation

Circumstance in which a person is exposed to hazard(s) (IEC 61508–4, 1998).

HAZOP

See Hazard and Operability Study.

HCI

See Human Computer Interface.

HECA

Hardware Error Confinement Area.

HHA

Health Hazard Assessment.

Hiding, Information

See Information Hiding.

HIFT

Hardware Implemented Fault Tolerance.

High-Level Requirements

Software requirements developed from analysis of system requirements, safety-related requirements, and system architecture (RTCA, DO178B, 1992).

HMI

See Human–Machine Interface.

Homogeneous Redundancy

In fault tolerance, realisation of the same function with identical means (IEEE Std–610.12, 1991).

Redundancy with the same kind of means. The means which are not the same can e.g. be: other physical principles, other ways of solving the same task, other procedures (VDE 0801, 1990).

Hot Restart

Restart of a system after power failure which occurs within the process dependent maximum time allowed for the system to recover as if there had been no power failure. All I/O information and other dynamic data as well as the application program context are restored or unchanged. Restart capability requires a separately powered real time clock or timer to determine elapsed time since the power failure was detected and a user accessible means to program the process dependent maximum time allowed (IEC 1131-1, 1992).

Hot Stand-by

A configuration in which a redundant functional unit can be immediately brought into service should the primary functional unit fail (ISO 2382-14, 1997).

Human-Computer Interface

The hardware and software that enables human beings to interact with the computer system (SSCP 85, 1995).

Human Error

Mistake. A human action, which can result in unintended system behaviour/failure (CENELEC, ENV50129, 1998).

Mistake. Human action or inaction that can produce an unintended result (IEC 61508-4, 1998; ISO 2382-14, 1997).

The failure of a person to perform an action as required (ESA, ECSS-P-001A, 1997).

A human action that produces an unintended and unwanted result (NATO, ARMP-7, 1996).

Mistakes by people, such as designers, engineers, operators, maintenance personnel or managers that may contribute to or result in undesired events (CCPS, 1993).

See Mistake (BSI, BS 4778-3.2, 1991; IEC 50-191, 1990).

Human–Machine Interface

The operators' windows to monitoring and keys, knobs switches, etc. for making adjustments in the process (CCPS, 1993).

See User Interface (IEEE Std–610.12, 1991).

HW

See Hardware.

I

Identification, Configuration

See Configuration Identification.

Idle

Pertaining to a system or component that is operational and in service, but not in use (IEEE Std–610.12, 1991).

IFR

Increasing Failure Rate.

IFT

Integral Functional Test.

Impact Analysis

Activity of identifying the effect that a change to a function or component in a system will have to other functions or components in that system as well as to other systems (IEC 61508–4, 1998).

Impairments to Dependability

Undesired, but not unexpected, circumstances causing or resulting from undependability. Faults, errors and failures (Laprie, 1992).

Implementation

A phase of the Development Process wherein the detailed specification of a Target of Evaluation is translated into actual hardware and software (UK DTI, ITSEC, 1991).

The process of translating a design into hardware components, software components, or both (IEEE Std-610.12, 1991).

Implementation Phase

The period of time in the software life cycle during which a software product is created from design documentation and debugged (IEEE Std-610.12, 1991; IEEE Std-1012, 1987).

Implementation Requirement

A requirement that specifies or constrains the coding or construction of a system or system component (IEEE Std-610.12, 1991).

Inactive Inventory

The group of items being held in reserve for possible future assignments to an operational status (USA DoD, MIL-Std-721C, 1992).

Inactive Time

That time during which an item is in an inactive or non-operational inventory, i.e. it is being held in reserve for potential future assignment to the inventory of an operational unit (NATO, ARMP-7, 1996).

That time during which an item is in reserve (in the inactive inventory) (USA DoD, MIL-Std-721C, 1992).

Incident

An unplanned event that could have been an accident, but was not (ESA, ECSS-P-001A, 1997).

The initial indication of a possible defect or failure (NATO, ARMP-7, 1996).

Any event indicating a possible non-conformance with the specification and which is reported on the R&M recording system (UK MoD, Def Stan 00-49, 1996).

An event that may result in a specification change (UK MoD, Def Stan 00-56, 1996).

A near miss or incident is an event that involves no loss (or only minor loss) but with the potential for loss under different circumstances (Leveson, 1995).

See Software Test Incident (IEEE Std-610.12, 1991; IEEE Std-1008, 1987).

An undesired event that, under slightly different circumstances, could have resulted in personal harm or property damage; any undesired loss of resources. Sometimes referred to as a near miss (ASSE, 1988).

Incident, Minor

See Minor Incident.

Incident, Software Test

See Software Test Incident.

Incipient Failure

An imperfection in the state or condition of an item or equipment so that a degraded or critical failure can be expected to result if corrective action is not taken (OREDA, 1992).

A failure that is about to occur (IEEE Std-610.12, 1991).

Inconsistent Failure

Failure such that system users may have different perceptions of it (Laprie, 1992).

Incorrect Service

Service delivered not in compliance with the system specification (Laprie, 1992).

Incredible

Believed to have a probability of occurrence too low for expression in meaningful numerical terms (UK MoD, Def Stan 00-56, 1996).

Independence

(Human) Freedom from intellectual, commercial and/or management involvement. (Technical) Freedom from any mechanism which can affect the correct operation of more than one item (CENELEC, ENV50129, 1998).

Separation of responsibilities which ensures the accomplishment of objective evaluation. (1) For software verification process activities, independence is achieved when the verification activity is performed by a person(s) other than the developer of the item being verified, and a tool(s) may be used to achieve an equivalence to the human verification activity. (2) For the software assurance process, independence also includes the authority to ensure corrective action (RTCA, DO178B, 1992).

Independent Failure

Failure which occurs without being caused by the failure of any other item (USA DoD, MIL-Std-721C, 1992).

Independent Fault

Faults attributed to different causes (Laprie, 1992).

Independent Function

Functions with immunity from common cause failure (UK MoD, Def Stan 00–56, 1996).

Independent Protection Layer

A system or subsystem specifically designed to reduce the likelihood of an identified hazard event by a large factor, i.e. at least by a 100-fold reduction in likelihood. An IPL must be independent of other protection layers associated with the identified hazard event, as well as dependable, and auditable (CCPS, 1993).

Independent Verification and Validation

Verification and validation performed by an organisation that is technically, managerially, and financially independent of the development organisation (IEEE Std–610.12, 1991).

Verification and validation performed by a contractor or government agency that is not responsible for developing the product or performing the activity being evaluated (USA DoD, MIL-Hdbk–347, 1990; USA DoD, MIL-Std–2167A, 1988).

Individual Risk

The frequency at which an individual may be expected to sustain a given level of harm from the realisation of specified hazards (Jones, 1992).

Infant Mortality

The set of failures that occur during the early-failure period of a system or component (IEEE Std–610.12, 1991).

Inference

The process of making a model specific, based on data that have been collected (Musa et al., 1987).

Informal Testing

Testing conducted in accordance with test plans and procedures that have not been reviewed and approved by a customer, user, or designated level of management (IEEE Std–610.12, 1991).

Information Hiding

A software design principle that the details of data representation should be concealed (SSP 84, 1996; SSCP 83, 1995).

A software development technique in which each module's interfaces reveal as little as possible about the module's inner workings and other modules are prevented from using information about the module that is not in the module's interface specification (IEEE Std–610.12, 1991).

Inherent Fault

Fault associated with original software product (whether new or modified) at completion or coding (Musa et al., 1987).

Inherited Code

Deliverable executable instructions that are reused from a previous application (this code is usually well debugged) (Musa et al., 1987).

Inherited Error

An error carried forward from a previous step in a sequential process (IEEE Std–610.12, 1991).

Inhibit

(1) To stop or prevent the occurrence of an event. (2) A design feature that provides a physical interruption between an energy source and a function actuator (ESA, ECSS-P–001A, 1997).

Initialisation

The process whereby a declared object is given some initial value. This may occur explicitly as a consequence of an initial value appearing in the declaration of the object or implicitly as a property of the type of the object (or both). Implicit initialisation may also perform other actions and is the opposite of finalisation (SSP 84, 1996).

Originating or establishing basic conditions or start-up state (IEC 1131-1, 1992; IEC 902, 1987).

Initiating Event

A postulated occurrence capable of leading to the realisation of a hazard (Jones, 1992).

Injury

Physical harm or damage to a person resulting from traumatic contact between the body and an outside agency or from exposure to environmental factors (ASSE, 1988).

Input

(1) The current, voltage, power or driving force applied to a circuit or device; also called input value or input data. (2) The terminals or other places where current, voltage, power or driving force may be applied to a circuit or device; also called input terminal. (3) Data which can be read by the application program from an interface between the user program execution function and another functional unit of the Programmable Controller System, as defined in this part of IEC 1131. Also called program input (IEC 1131-1, 1992).

(1) Pertaining to data received from an external source. (2) Pertaining to a device, process, or channel involved in receiving data from an external source. (3) To receive data from an external source. (4) Loosely, input data (IEEE Std-610.12, 1991).

Input Module

E/E/PES or subsystem that acts as an interface to external devices and converts input signals into signals that the E/E/PES can utilise (ISA, S84.01, 1996).

Input/Output Station, Remote

See Remote Input/Output Station.

Input Space

The set of all possible input states for a program (Musa et al., 1987).

Input State

Set of input variables for a particular run (Musa et al., 1987).

Input Variable

Any data item that exists external to a run and is used by the run (Musa et al., 1987).

Inspection

Activity such as measuring, examining, testing or gauging one or more characteristics of an entity and comparing the results with specified requirements in order to establish whether conformity is achieved for each characteristic (ESA, ECSS-P-001A, 1997; ISO 8402, 1994).

Conformity evaluation by observation and judgement accompanied as appropriate by measurement, testing or gauging (ISO/IEC, Guide 2, 1996).

The examination and testing of supplies and services (including, when appropriate, raw materials, components, and intermediate assemblies) to determine whether they conform to specified requirements (USA DoD, MIL-Std-109C, 1994).

Activities such as measuring, examining, testing, gauging one or more characteristics of a product or service and comparing these with specified requirements to determine conformity (BSI, BS 4778-2, 1991; BSI, BS 4778-1, 1991).

A static analysis technique that relies on visual examination of development products to detect errors, violations of development standards, and other problems (IEEE Std-610.12, 1991).

Installation and Checkout Phase

The period of time in the software life cycle during which a software product is integrated into its operational environment and tested in this environment to ensure that it performs as required (IEEE Std–610.12, 1991; IEEE Std–1012, 1987; IEEE Std–729, 1983).

Installation, Permanent

See Permanent Installation.

Instruction

A programming language element that specifies an operation and the values or location of its operands (IEC 1131–1, 1992).

A statement that specifies an operation and the values or locations of its operands (IEC 902, 1987).

Instruction List Language

A textual programming language using instructions for representing the application program for a Programmable Controller System (IEC 1131–1, 1992).

Instrumented Protective Function

A function composed of sensors, logic solver and final control elements for the purpose of preventing Hazards (Korteweg, 1994).

Instrumented Protective System

The (electrical and/or electronic and/or programmable electronic) logic solver component of the Instrumented Protective Function (Korteweg, 1994).

Instrumented System, Safety

See Safety Instrumented System.

Integration

Process of assembling multiple components or subsystems to form a system (ISA, S84.01, 1996).

The prevention of the unauthorised modification of information (UK DTI, ITSEC, 1991).

The process of combining software components, hardware components, or both into an overall system (IEEE Std–610.12, 1991).

The step-by-step process of testing where each module is tested alone and then in conjunction with others until the system is built up (Smith & Wood, 1989).

Integration, Hardware/Software

See Hardware/Software Integration.

Integration, Software

See Software Integration.

Integration Testing

Testing in which software components, hardware components, or both are combined and tested to evaluate the interaction between them (IEEE Std–610.12, 1991).

An orderly progression of testing in which software elements, hardware elements, or both are combined and tested until the entire system has been integrated (IEEE Std–1012, 1987; IEEE Std–729, 1983).

Integrity

Condition of being unimpaired (Laprie, 1992).

The degree to which a system or component prevents unauthorised access to, or modification of, computer programs or data (IEEE Std–610.12, 1991).

The ability of a system to perform its functions correctly when required. The term 'integrity' is usually associated with safe systems (Smith & Wood, 1989).

The data quality that exists as long as accidental or malicious destruction, alteration, or loss of data does not occur (ISO/IEC 2382-8, 1986).

The extent to which unauthorised access to or modification of the product or data can be controlled in a (computer) system (IEEE Std-729, 1983).

Integrity, Conventional Safety

See Conventional Safety Integrity.

Integrity Level

An indicator of SIS performance as measured by PFD (i.e. 1-availability) (CCPS, 1993).

Integrity Level, Safety

See Safety Integrity Level.

Integrity, Random Failure

See Random Failure Integrity.

Integrity, Safety

See Safety Integrity.

Integrity, Systematic Failure

See Systematic Failure Integrity.

Intensity, Failure

See Failure Intensity.

Intent, Design

See Design Intent.

Intentional Fault

Fault created deliberately (Laprie, 1992).

Interchange

Removing the item that is to be replaced, and installing the replacement item (USA DoD, MIL-Std-721C, 1992).

Interchangeable Item

One, which (1) possesses such functional and physical characteristics as to be equivalent in performance, reliability, and maintainability, to another item of similar or identical purposes; and (2) is capable of being exchanged for the other item (a) without selection for fit or performance, and (b) without alteration of the items themselves or of adjoining items, except for adjustment (USA DoD, MIL-Hdbk-505, 1998; USA DoD, MIL-Std-280A, 1969).

Interconnection

The link between two components (at whatever level) within a system, which exists because of (and is defined by) an interaction between the components. This link may be logical or physical (UK MoD, Def Stan 00-58, 1996).

Interface

A shared boundary involving the specification of the interconnection between two equipments or systems. The specification includes the type, quantity and function of the interconnecting elements and the type and form of function or information to be interchanged between them (UK MoD, Def Stan 00-49, 1996).

Shared boundary through which information is conveyed (ISA, S84.01, 1996).

A shared boundary between two functional units, defined by various characteristics pertaining to the functions, physical interconnections, signal exchanges, and other characteristics as appropriate (ISO 2382–1, 1993).

The functional and physical characteristics required to exist at a common boundary (USA DoD, MIL-Std–973, 1992).

(1) A shared boundary across which information is passed. (2) A hardware or software component that connects two or more other components for the purpose of passing information from one to the other. (3) To connect two or more components for the purpose of passing information from one to the other. (4) To serve as a connecting or connected component as in (2) (IEEE Std–610.12, 1991).

Interface, Graphical User

See Graphical User Interface.

Interface, Human–Computer

See Human–Computer Interface.

Interface, Human–Machine

See Human–Machine Interface.

Interface, Man–Machine

See Man–Machine Interface.

Interface Testing

Testing conducted to evaluate whether systems or components pass data and control correctly to one another (IEEE Std–610.12, 1991).

Interference, Electromagnetic

See Electromagnetic Interference.

Interlock System

A system that, in response to a predetermined condition, initiates or prevents a predefined action (CCPS, 1993).

Intermittent Failure

Failure for a limited period of time, followed by the item's recovery of its ability to perform within specified limits without any remedial action (USA DoD, MIL-Std-721C, 1992).

Intermittent Fault

Temporary internal fault. Faults whose conditions of activation cannot be reproduced or which occur rarely enough (Laprie, 1992).

A temporary or unpredictable fault in a component (IEEE Std-610.12, 1991).

Internal Communication

Data exchange between the various devices within a given Safety Instrumented System. These include bus backplane connections, the local or remote I/O bus, etc. (ISA, S84.01, 1996).

Internal Failure

A failure involving the manufacturer-supplied hardware or software of the Programmable Controller System, excluding the application program (IEC 1131-1, 1992).

Internal Fault

Fault inside a system (Laprie, 1992).

Interrupt

A break in the normal flow of a system or program occurring in such a way that the flow can be resumed from that point at a later time (IEC 1131-1, 1992; IEC 902, 1987).

A suspension of a task, such as the execution of a computer program, caused by an event external to that task, and performed in such a way that task can be resumed (RTCA, DO178B, 1992).

(1) The suspension of a process to handle an event external to the process. (2) To cause the suspension of a process. (3) Loosely, an interrupt request (IEEE Std-610.12, 1991).

Interval, Test

See Test Interval.

Intrinsic Availability

The probability that the system/equipment is operating satisfactorily at any point in time when used under stated conditions, where the time considered is operating time and repair time (active) (NATO, ARMP-7, 1996).

The probability that the system is operating satisfactorily at any point in time when operated under specified conditions, where the time categories considered are only operating time and active repair time (Martz & Waller, 1982).

Intrinsic Reliability

The basic reliability level indicated by the design and components and taking into account failures added by manufacture, installation, wearout or operation (Smith, 1981).

Intrinsic Safety

A degree of integrity designed into a system in order to meet defined safety criteria (Smith & Wood, 1989).

Intrusion

Intentional operational external fault (Laprie, 1992).

Inventory, Active

See Active Inventory.

Inventory, Inactive

See Inactive Inventory.

Involved, Safety

See Safety Involved.

IPF

See Instrumented Protective Function.

IPL

See Independent Protection Layer.

IRS

Interface Requirements Specifications.

Isolated

Devices, circuits are said to be isolated when there is no galvanic connection between them (IEC 1131-1, 1992).

Isolation, Electrical

See Electrical Isolation.

Isolation, Fault

See Fault Isolation.

IT

Information Technology.

Item

Element under consideration (CENELEC, ENV50129, 1998).

A non-specific term used to denote any product, including systems, materials, parts, subassemblies, sets, accessories, etc. (USA DoD, MIL-Hdbk-505, 1998; USA DoD, MIL-Std-721C, 1992; USA DoD, MIL-Std-280A, 1969).

Anything which can be individually described and considered (ESA, ECSS-P-001A, 1997; ISO 8402, 1994).

The component or set of components within the boundary specified for the particular item (OREDA, 1992).

Any part, component, device, subsystem, functional unit, equipment or system that can be individually considered (BSI, BS 4778-3.2, 1991; IEC 50-191, 1990).

Item, Configuration

See Configuration Item.

Item, Critical

See Critical Item.

Item, Developmental

See Developmental Item.

Item, Interchangeable

See Interchangeable Item.

Item, Off-the-Shelf

See Off-the-Shelf Item.

Item, Reparable

See Reparable Item.

Item, Replacement

See Replacement Item.

Item, Smallest Replaceable

See Smallest Replaceable Item.

Item, Software

See Software Item.

Item, Software Configuration

See Software Configuration Item.

Item, Substitute

See Substitute Item.

Item, Test

See Test Item.

ITSEC

Information Technology Security Evaluation Criteria.

ITSEM

Information Technology Security Evaluation Manual.

IV&V

See Independent Verification and Validation.

J

Jump, Conditional

See Conditional Jump.

Jump, Unconditional

See Unconditional Jump.

K

Kernel, Security

See Security Kernel.

Kind, Replacement in

See Replacement in Kind.

KIS

Keep It Simple.

Knock, Double

See Double Knock.

L

Ladder Diagram

One or more networks of contacts, coils, graphically represented functions, function blocks, data elements, labels, and connective elements, delimited on the left and (optionally) on the right by power rails (IEC 1131-1, 1992).

Ladder Diagram Language

A programming language using ladder diagrams for representing the application program for a Programmable Controller System (IEC 1131-1, 1992).

LAN

Local Area Network.

Language, Application

See Application Language.

Language, Assembly

See Assembly Language.

Language, Block Structured

See Block Structured Language.

Language, Formal

See Formal Language.

Language, Function Block Diagram

See Function Block Diagram Language.

Language, Graphical

See Graphical Language.

Language, Instruction List

See Instruction List Language.

Language, Ladder Diagram

See Ladder Diagram Language.

Language, Limited Variability

See Limited Variability Language.

Language, Logic Programming

See Logic Programming Language.

Language, Machine

See Machine Language.

Language, Programming

See Programming Language.

Language, Structured Text

See Structured Text Language.

Language, Textual

See Textual Language.

Latency

Hazard exposure or duration (Leveson, 1995).

Latency Period

The interval of time between the occurrence of a fault and its detection or its resulting outcome (Hölscher & Rader, 1986).

Latent Error

Error not recognised as such (Laprie, 1992).

Latent Fault

An existing fault that has not yet been recognised (BSI, BS 4778-3.2, 1991; IEC 50-191, 1990).

LCC

Life Cycle Cost.

LCSAJ

See Linear Code Sequence and Jump.

Least Replaceable

That assembly at which diagnosis ceases and replacement is carried out (typical for printed board assembly) (Smith, 1981).

Library

A collection of software usually written by a third party and usually provided in binary form for incorporation into a system (SSP 84, 1996).

Library, Software

See Software Library.

Life Cycle

Time interval between a product's conception and its disposal (IEC 300-3-3, 1996).

A framework containing the processes, activities and tasks involved in the design, development, operation and maintenance of a software product, spanning the life of the system from the definition of its requirements to the termination of its use (UK MoD, Def Stan 05-95, 1995).

The time interval from a product's inception until its ultimate disposal and removal from use (IEC 300-2, 1995).

A generic term covering all phases of acquisition, operation, and logistics support of an item, beginning with concept definition and continuing through disposal of the item (USA DoD, MIL-Std-973, 1992).

Life Cycle Phase

Any period of time during software development or operation that may be characterised by a primary type of activity (such as design or testing) that is being conducted. These phases may overlap one another; for V&V purposes, no phase is concluded until its development products are fully verified (IEEE Std–1012, 1987).

Life Cycle Profile

A thorough time-life description of the events and conditions associated with an item of equipment from the time of final factory acceptance until its ultimate disposition. Each significant life cycle event, such as transportation, dormant storage, test and checkout, stand-by and ready modes, operational deployment, and mission profiles, is addressed, including alternative possibilities. The profile depicts the time span of each event, the environmental conditions, and the operating modes (USA DoD, MIL-Hdbk–781A, 1996).

Life Cycle, Safety

See Safety Life Cycle.

Life Cycle, Software

See Software Life Cycle.

Life Cycle, System

See System Life Cycle.

Life, Observed B-Percentile

See Observed B-Percentile Life.

Life Profile

A time-phased description of the events and environments an item experiences from manufacture to final expenditures or removal from the operational inventory, to include one or more mission profiles (USA DoD, MIL-Std–721C, 1992).

Life, Shelf

See Shelf Life (USA DoD, MIL-Std-721C, 1992).

Life, Storage

See Storage Life.

Life Units

A measure of use duration applicable to the item (e.g. operating hours, cycles, distance, rounds fired, attempts to operate, etc.) (USA DoD, MIL-Std-721C, 1992).

Life, Useful

See Useful Life.

Limited Variability Language

Software programming language, either textual or graphical, for commercial and industrial programmable electronic controllers with a range of capabilities limited to their application (IEC 61508-4, 1998).

Limiting Resource

The resource that controls the relationship between execution time and calendar time (Musa et al., 1987).

Line

See Train (ISA, S88.01, 1995).

Linear Code Sequence and Jump

A control flow sequence consisting of a series of sequential instructions, followed by a branch to a non-sequential instruction (UK MoD, Def Stan 00-55, 1997).

Load Test

A system test involving simulated inputs in order to prove that the system will function at full load (Smith, 1981).

Localisation, Fault

See Fault Localisation.

Lock-Out, Non-Volatile

See Non-Volatile Lock-Out.

Lock-Out, Volatile

See Volatile Lock-Out.

Log, Test

See Test Log.

Logic Circuit

Interconnected electrical, electronic, and micro-electronic circuits that carry and control electrical impulses, for the purpose of providing the 'ON-OFF' intelligence for predetermined relationships within a system or subsystem (USA DoD, MIL-Hdbk-272A, 1993).

Logic Control System

A control system in which definite output signal states are functions of the states of the input signals in keeping with the rules of Boolean algebra (IEC 1131-1, 1992).

Logic, Malicious

See Malicious Logic.

Logic Programming Language

A programming language used to express programs in terms of control constructs and a restricted predicate calculus; e.g. PROLOG (IEEE Std–610.12, 1991).

Logic Solver

E/E/PES components or subsystems that execute the application logic. Electronic and programmable electronics include input/output modules (ISA, S84.01, 1996).

The portion of the SIS that performs the logic function (e.g. electromechanical relays, the CPU of a PES, etc.) (CCPS, 1993).

Logic System

Portion of a system that performs the function logic but excludes the sensors and final elements (IEC 61508–4, 1998).

Logistic Support

The overall resources which are arranged and organised in order to operate and maintain the system at the specified availability level at the required life cycle cost (CENELEC, prEN50126, 1998).

Logistic Time

The portion of down-time spent waiting for replacement parts (Martz & Waller, 1982).

Loop Control, Open

See Open Loop Control.

Loopback Testing

Testing in which signals or data from a test device are input to a system or component, and results are returned to the test device for measurement or comparison (IEEE Std–610.12, 1991).

Low-Level Requirements

Software requirements derived from high-level requirements, derived requirements, and design constraints from which source code can be directly implemented without further information (RTCA, DO178B, 1992).

M

Machine Language

An artificial language composed only of the machine instructions of a specific computer or class of computers (ISO/IEC DIS 2382-7, 1996).

Main Processing Unit

The portion of the Programmable Controller System which interprets or executes the (main part of the) application program. The MPU may include power supply, memory, and I/Os (IEC 1131-1, 1992).

Maintainability

The probability that a given active maintenance action, for an item under given conditions of use can be carried out within a stated interval time interval when the maintenance is performed under stated conditions and using stated procedures and resources (CENELEC, ENV50129, 1998; CENELEC, prEN50126, 1998; NATO, ARMP-7, 1996).

The ability of a functional unit, under given conditions of use, to be retained in, or restored to, a state in which it can perform a required function when maintenance is performed under given conditions and using stated procedures and resources (ISO 2382-14, 1997).

The ability of an item under given conditions of use, to be retained in, or restored to, a state in which it can perform a required function, when maintenance is performed under given conditions and using stated procedures and resources (ESA, ECSS-P-001A, 1997; BSI, BS 4778-3.2, 1991; IEC 50(191), 1990).

The ease with which a software system or component can be modified to correct faults, improve performance or other attributes or adapt to a changed environment (UK MoD, Def Stan 00-55, 1997).

A characteristic of design and installation which is expressed as the probability that an item will be retained in or restored to a specified condition within a given period of time, when the maintenance is performed in accordance with prescribed procedures and resources (USA DoD, MIL-Std-109C, 1994).

Easiness with which maintenance actions can be performed. Measure of continuous incorrect service delivery. Measure of the time to restoration from the last experienced failure (Laprie, 1992).

(1) The ease with which a software system or component can be modified to correct faults, improve performance or other attributes, or adapt to a changed environment. (2) The ease with which a hardware system or component can be retained in, or restored to, a state in which it can perform its required functions (IEEE Std-610.12, 1991).

The probability that a failed system can be restored to operating conditions in a specified interval of down-time, which excludes only any free down-time (Martz & Waller, 1982).

The ability of an item, under stated conditions of use, to be retained in, or restored to, a state in which it can perform its required functions, when maintenance is performed under stated conditions and using prescribed procedures and resources (O'Connor, 1981).

The probability that a failed item will be restored to operational effectiveness within a given period of time when the repair is performed in accordance with prescribed procedures (Smith, 1981).

Maintainability, Mission

See Mission Maintainability.

Maintenance

The combination of all technical and administrative actions, including supervision actions, intended to retain a product in, or restore it to, a state in which it can perform a required function (CENELEC, ENV50129, 1998; CENELEC, prEN50126, 1998; ESA, ECSS-P-001A, 1997; BSI, BS 4778-3.2, 1991; IEC 50-191, 1990).

A set of activities intended to keep a functional unit in, or to restore it to, a state in which it can perform a required function (ISO 2382-14, 1997).

(1) All action taken to retain materiel in or to restore it to a specified condition. It includes inspection, testing, servicing, classification as serviceability, repair, rebuilding and reclamation. (2) All supply and repair action taken to keep a force in condition to carry out its mission. (3) The routine recurring work required to keep a facility (plant, building, structure, ground facility, utility system or other real property) in such condition that it may be continuously utilised, at its original or designed capacity and efficiency for its intended purpose (NATO, ARMP-7, 1996).

All action taken to retain materiel in, or restore it to, a specified condition. It includes: inspection, testing, servicing, classification as to serviceability, repair, rebuilding, and reclamation (UK MoD, Def Stan 00-49, 1996).

All actions necessary for retaining an item in or restoring it to a specified condition (USA DoD, MIL-Std-721C, 1992).

(1) The process of modifying a software system or component after delivery to correct faults, improve performance or other attributes, or adapt to a changed environment. (2) The process of retaining a hardware system or component in, or restoring it to, a state in which it can perform its required functions (IEEE Std-610.12, 1991).

Any activity intended to retain a functional unit in, or to restore it to, a state in which it can perform its required function (IEC 902, 1987).

Maintenance Action

An element of a maintenance event. One or more tasks (i.e., fault localisation, fault isolation, servicing and inspection) necessary to retain an item in or restore it to a specified condition (USA DoD, MIL-Std-721C, 1992).

Maintenance, Adaptive

See Adaptive Maintenance.

Maintenance, Corrective

See Corrective Maintenance.

Maintenance, Emergency

See Emergency Maintenance.

Maintenance, Event

See Event Maintenance.

Maintenance, Perfective

See Perfective Maintenance.

Maintenance Phase, Operation and

See Operation and Maintenance Phase.

Maintenance Policy

A description of the interrelationship between the maintenance echelons, the indenture levels and the levels of maintenance to be applied for the maintenance of an item (CENELEC, prEN50126, 1998; BSI, BS 4778-3.2, 1991; IEC 50-191, 1990).

Maintenance, Preventive

See Preventive Maintenance.

Maintenance Ratio

A measure of the total maintenance manpower burden required to maintain an item. It is expressed as the cumulative number of man-hours of maintenance expended in direct labour during a given period of the life units divided by the cumulative number of end item life units during the same period (USA DoD, MIL-Std-721C, 1992).

Maintenance, Scheduled

See Scheduled Maintenance.

Maintenance, Software

See Software Maintenance.

Maintenance Time

An element of down-time which excludes modification and delay time (USA DoD, MIL-Std-721C, 1992).

The time interval during which a maintenance action is performed on an item either manually or automatically, including technical delays and logistic delays (BSI, BS 4778-3.2, 1991; IEC 50-191, 1990).

Maintenance, Unscheduled

See Unscheduled Maintenance.

Malfunction

See Fault (NATO, ARMP-7, 1996).

See Failure (USA DoD, MIL-Std-721C, 1992).

Malicious Logic

Intentional design fault (Laprie, 1992).

Man-Machine Interface

A manufacturer's catalogued peripheral equipped with pushbuttons, lamps, keyboards, displays or equivalent, intended as operator interface, such as motor control/monitor panel, general-purpose operator interface, etc. (IEC 1131-1, 1992).

Management, Configuration

See Configuration Management.

Management, Quality

See Quality Management.

Management, Risk

See Risk Management.

Management, Safety

See Safety Management.

Management, Software Reliability

See Software Reliability Management.

Manufacturing Phase

The period of time in the software life cycle during which the basic version of a software product is adapted to a specified set of operational environments and is distributed to a customer base (IEEE Std–610.12, 1991; IEEE Std–1002, 1987).

Masking, Fault

See Fault Masking.

Materiel

A collective term for the articles, supplies, machinery, etc. used in an organisation, as distinguished from the personnel or body of persons employed (ESA, ECSS-P-001A, 1997).

A generic term encompassing equipment, stores, supplies and spares (UK MoD, Def Stan 00–49, 1996).

A generic term covering systems, equipment, stores, supplies and spares and including related documentation manuals, computer software and firmware (UK MoD, Def Stan 05–57, 1993).

MDT

Mean Down-Time.

Mean Time Between Failure

The mean duration between consecutive failures of a functional unit under given conditions (ISO 2382–14, 1997).

The expectation of the operating time between failures (ESA, ECSS-P-001A, 1997; BSI, BS 4778-3.2, 1991; IEC 50–191, 1990).

A basic measure of the system reliability parameter related to availability and readiness. The total number of system life units, divided by the total number of events in which the system becomes unavailable to initiate its mission(s) during a stated period of time (USA DoD, MIL-Hdbk-781A, 1996).

A measure of the expected (average) time during which a system will continuously perform within its specified limits under stated conditions (NATO, ARMP-7, 1996).

A basic measure of reliability for reparable items. The mean number of life units during which all parts of the item perform within their specified limits, during a particular measurement interval under stated conditions (USA DoD, MIL-Std-109C, 1994; USA DoD, MIL-Std-721C, 1992).

For a stated period in the life of a functional unit, the mean value of the length of time between consecutive failures under stated conditions (IEC 1131-1, 1992; IEC 902, 1987).

The expected or observed time between consecutive failures in a system or component (IEEE Std-610.12, 1991).

The total cumulative functioning time of a population divided by the number of failures. As with failure rate the same applies to observed, assessed and extrapolated MTBF. MTBF is used for items which involve repair. MTTF is used for items with no repair (Smith, 1981).

Mean Time Between Failure, Observed

See Observed Mean Time Between Failure.

Mean Time To Failure

This is similar to Mean Time Between Failure but applies to non-repairable systems (NATO, ARMP-7, 1996).

A basic measure of reliability for non-repairable items: The total number of life units of an item divided by the total number of failures within that population, during a particular measurement interval under stated conditions (USA DoD, MIL-Std-721C, 1992).

The expectation of the time to failure (BSI, BS 4778-3.2, 1991; IEC 50-191, 1990).

The average time the system or component of the system works without failing (IEC 902, 1987).

Expected value of the failure interval (Musa et al., 1987).

Mean Time To Repair

The average or expected time it takes to repair an equipment (NATO, ARMP-7, 1996).

A basic measure of maintainability: The sum of corrective maintenance times at any specific level of repair, divided by the total number of failures within an item repaired at that level, during a particular interval under stated conditions (USA DoD, MIL-Std-721C, 1992).

The expected or observed time required to repair a system or component and return it to normal operations (IEEE Std-610.12, 1991).

The average time required for corrective maintenance (IEC 902, 1987).

Expected value of the time required to restore to normal operation (Musa et al., 1987).

The mean time to carry out a defined maintenance action (usually refers to corrective maintenance) (Smith, 1981).

Mechanism, Failure

See Failure Mechanism.

Media

Devices or material which act as a means of transferral or storage of software, for example, programmable ROM, magnetic tapes or disks, and paper (RTCA, DO178B, 1992).

Media, Software

See Software Media.

Memory Device

An article of hardware capable of storing machine-readable computer programs and associated data. It may be an integrated circuit chip, a circuit card containing integrated circuit chips, a core memory, a disk, or a magnetic tape (RTCA, DO178B, 1992).

Memory, User Program

See User Program Memory.

Memory Utilisation

The amount of memory required for a specified part of a program (IEC 1131-1, 1992).

Memory, Volatile

See Volatile Memory.

Metric

See Software Quality Metric (IEEE Std–1061, 1992).

A quantitative measure of the degree to which a system, component, or process possesses a given attribute (IEEE Std–610.12, 1991).

Parameters for describing the structure, size and complexity of a program. Attempts are made to relate these metrics, by regression techniques, to software quality (Smith & Wood, 1989).

Metric, Direct

See Direct Metric.

Metric, Predictive

See Predictive Metric.

Metric, Process

See Process Metric.

Metric, Product

See Product Metric.

Metric, Software Quality

See Software Quality Metric.

Metric, Validated

See Validated Metric.

Metrix, Complexity

See Complexity Metrix.

Microcode

A set of sequences of elementary instructions (lower level than machine language) that correspond to computer operations, that are maintained in special storage, and whose execution is initiated by the introduction of a computer instruction in to an instruction register of a computer. Microcode is often used in place of hard-wired logic, and is usually inaccessible in programming (USA DoD, MIL-Hdbk-272A, 1993).

Minor Incident

Impact initially limited to local area of the event with potential for broader consequence if corrective action is not taken.

Mishap

An unplanned event or series of events resulting in death, injury, occupational illness, or damage to or loss of equipment or property, or damage to the environment. Accident (USA DoD, MIL-Std-882C, 1993).

Mission

An objective description of the fundamental task performed by a system (CENELEC, prEN50126, 1998).

The specific task, duty or function defined to be accomplished by a system (ESA, ECSS-P-001A, 1997).

Mission Maintainability

The probability that an item will perform its required functions for the duration of a specified mission profile (NATO, ARMP-7, 1996).

The measure of the ability of an item to be retained in or restored to specified condition when maintenance is performed during the course of a specified mission profile (the mission-related system maintainability parameter) (USA DoD, MIL-Std-721C, 1992).

The ability of an item to perform its required functions for the duration of a specified mission profile (USA DoD, MIL-Std-785B, 1988).

Mission Profile

Outline of the expected range and variation in the mission with respect to parameters such as time, loading, speed, distance, stops, tunnels, etc., in the operational phases of the life cycle (CENELEC, prEN50126, 1998).

A thorough description of all of the major planned events and conditions associated with one specific mission. A mission profile is one segment of a life cycle profile. The profile depicts the time span of the event, the expected environmental conditions, energised and non-energised periods, and so forth (USA DoD, MIL-Hdbk-781A, 1996).

A time-phased description of the events and environments an item experiences from initiation to completion of a specified mission, to include the criteria of mission success or critical failures (USA DoD, MIL-Std-721C, 1992).

Mission Reliability

The ability of an item to perform its required functions for the duration of a specified 'mission profile' (USA DoD, MIL-Std-721C, 1992).

Mission Time

That element of up-time required to perform a stated mission profile (NATO, ARMP-7, 1996; USA DoD, MIL-Std-721C, 1992).

Mistake

See Human Error (IEC 61508-4, 1998; CENELEC, ENV50129, 1998).

A human action that produces an incorrect result (IEEE Std-610.12, 1991).

Human error. A human action that produces an unintended result (BSI, BS 4778-3.2, 1991; IEC 50-191, 1990).

Human action which can lead to (but need not lead to) some undesirable consequence (Hölscher & Rader, 1986).

Mitigation

Makes a consequence less severe or relieves consequences (Korteweg, 1994).

Mitigation System

A system designed to respond to a failure event sequence by reducing its impact consequences (CCPS, 1993).

ML

Maximum Likelihood.

MLE

Maximum Likelihood Estimate.

MMI

Man–Machine Interface.

MOC

Management of Change.

Mode

The manner in which the transition of sequential functions are carried out within a procedural element or the accessibility for manipulating the states of equipment entities manually or by other types of control (ISA, S88.01, 1995).

Mode, Continuous

See Continuous Mode.

Mode, Demand

See Demand Mode.

Model, Error Prediction

See Error Prediction Model.

Modifiable

Information is modifiable if it is structured and has a style such that changes can be made completely, consistently, and correctly while retaining the structure (RTCA, DO178B, 1992).

Modification Request

A generic term that includes the forms associated with the various trouble/reporting documents and the configuration change control documents (IEEE Std–1219, 1992).

Modification Time

The time necessary to introduce any specific change(s) to an item to improve its characteristics or add new ones (USA DoD, MIL-Std–721C, 1992).

Modular Programming

A software development technique in which software is developed as a collection of modules (ISO/IEC DIS 2382–7, 1996).

Modularity

The property of functional flexibility built into a system by assembling discrete units which can be easily joined to or arranged with other parts or units (CENELEC, ENV50129, 1998).

A measure of the extent to which a program is composed of modules, such that a change to one module has minimal impact on other modules (ISO/IEC DIS 2382–7, 1996).

The degree to which a system or computer program is composed of discrete components such that a change to one component has minimal impact on other components (IEEE Std–610.12, 1991).

Module

An identifiable, self-contained, compilable, independent set of computer instructions with one entry point and one exit point that fulfils some well defined function (USA DoD, MIL-Hdbk–272A, 1993).

Module, Input

See Input Module.

Module, Output

See Output Module.

Module, Software

See Software Module.

Monitoring

Observation of the operation of a system or part of a system to verify correct functioning or detect incorrect functioning by measuring system variables and comparing the measured values with specified values (IEC 50–351, 1998).

Functionality within a system which is designed to detect anomalous behaviour of the system (RTCA, DO178B, 1992).

Activity, performed either manually or automatically, intended to observe the state of an item (BSI, BS 4778–3.2, 1991; IEC 50–191, 1990).

Observation of the operation of a system or part of a system to verify correct functioning and detect incorrect functioning, this being done by measuring one or more variables of the system and comparing the measured values with the specified values (IEC 902, 1987).

Those tests which run in parallel with useful operation (Hölscher & Rader, 1986).

Monitoring, Condition

See Condition Monitoring.

Monte Carlo Computer Simulation Technique

A method utilising random sampling to obtain inputs for computer simulation trials and obtaining approximate solutions in terms of a range of values each of which has a calculated probability of being the solution to the problem (NATO, ARMP-7, 1996).

MORT

Management Oversight and Risk Tree.

Mortality

The nature of humans as having eventually to die; the proportion of deaths in a given population or from a particular cause (ASSE, 1988).

Mortality, Infant

See Infant Mortality.

MPU

See Main Processing Unit.

MR

See Modification Request.

MTBF

See Mean Time Between Failure.

MTDF

Mean Time to Detect Failure.

MTTF

Mean Time To Failure.

MTTR

See Mean Time To Repair.

Multiple Failures

The simultaneous occurrence of two or more independent failures. When two or more failed parts are found during troubleshooting and failures cannot be shown to be dependent, multiple failures are presumed to have occurred (USA DoD, MIL-Hdbk-781A, 1996).

Mutation Testing

A testing methodology in which two or more program mutations are executed using the same test cases to evaluate the ability of the test cases to detect differences in the mutations (IEEE Std-610.12, 1991).

N

NDFIT

Non-Destructive Fault Insertion Testing.

NDI

See Non-Developmental Item.

Near Accident

In occupational safety, an event having a potential for personal injury, death, or property damage, but where there was no such consequence (ASSE, 1988).

Near Miss

See Incident (Leveson, 1995).

Negation

Enforcement of a safe state following detection of a hazardous fault (CENELEC, ENV50129, 1998).

NHPP

Non-Homogeneous Poisson Process.

Non-Chargeable Failure

(1) A non-relevant failure. (2) A relevant failure caused by a condition previously specified as not within the responsibility of a given organisational entity. (All relevant failures are chargeable to one organisational entity or another.) (USA DoD, MIL-Std-721C, 1992).

Non-Conformance

The failure of a characteristic to conform to the requirements specified in the contract, drawings, specifications, or other approved product descriptions (USA DoD, MIL-Std-109C, 1994).

The failure of a unit or product to conform to specified requirements (USA DoD, MIL-Std-973, 1992).

Non-Conformity

The non-fulfilment of specified requirements (BSI, BS 4778-1, 1991).

The non-fulfilment of intended usage requirements (IEC 902, 1987).

Non-Destructive Test

A test to determine the characteristics or properties of a material or substance that does not involve its destruction or detoriation (ASSE, 1988).

Non-Developmental Item

An item or supply that is either in the process of becoming or already is commercially available or is in use by a UK government agency or department or foreign government with which the UK has a mutual defence co-operation agreement (UK MoD, Def Stan 00-56, 1996).

(1) Any item of supply that is available in the commercial marketplace; (2) Any previously developed item of supply that is in use by a department or agency of the United States, a State or local government, or a foreign government with which the United States has a mutual defence co-operation agreement; (3) Any item of supply described in definition (1) or (2), above, that requires only minor modification in order to meet the requirements of the procuring agency; or (4) Any item of supply that is currently being produced that does not meet the requirements of definition (USA DoD, MIL-Std-882C, 1993).

Non-Relevant Failure

(1) A failure verified as having been caused by a condition not present in the operational environment. (2) A failure verified as peculiar to an item design that will not enter the operational inventory (USA DoD, MIL-Std-721C, 1992).

Non-Volatile Lock-Out

The safety shut-down condition of the system, such that a restart can only be accomplished by a manual reset of the system and by no other means (CEN, EN 298, 1993).

NOR

Notice of Revision.

Not Operating

The state wherein an item is able to function but is not required to function. Not to be confused with down-time (USA DoD, MIL-Std–721C, 1992).

Not Operating Time

That element of up-time during which the item is not required to operate (USA DoD, MIL-Std–721C, 1992).

NP

Non-Programmable.

Nuisance Trip

See Spurious Trip (ISA, S84.01, 1996).

NVP

N-Version Programming.

O

Object Code

A low-level representation of the computer program not usually in a form directly usable by the target computer but in a form which includes relocation information in addition to the processor instruction information (RTCA, DO178B, 1992).

Computer instructions and data definitions in a form output by an assembler or compiler. An object program is made up of object code (IEEE Std–610.12, 1991).

Objective, Test

See Test Objective.

Obligation, Proof

See Proof Obligation.

Observed B-Percentile Life

The length of observed time at which stated proportion (B%) of a sample of items has failed (O'Connor, 1981).

Observed Failure Rate

For a stated period in the life of an item, the ratio of the total number of failures in a sample to the cumulative observed time on that sample. The observed failure rate is to be associated with particular and stated time intervals (or summation of intervals) in the life of the item, and under stated conditions (O'Connor, 1981).

Observed Mean Time Between Failure

For a stated period in the life of an item, the mean value of the length of time between consecutive failures computed as the ratio of the cumulative observed time to the number of failures under stated conditions (O'Connor, 1981).

Off-Delay Timer

A function block delaying a Boolean input for a specified duration when changing from one to zero (IEC 1131-1, 1992).

Off-Line

Process, to which the Safety Instrumented System is connected, is shut down (ISA, S84.01, 1996).

Pertaining to a device or process that is not under the direct control of the central processing unit of a computer (IEEE Std-610.12, 1991).

When equipment or devices operate off-line to an industrial process, they are associated with this industrial process, but do not act on the process or are not acted upon by this process during the actual time in which the process takes place (IEC 902, 1987).

Off-Line Test

An off-line test is a test which is carried out outside the on-line operation, e.g. in corrective maintenance pauses and maintenance pauses (VDE 0801, 1990).

Off-the-Shelf Equipment

Equipment that is available for acquisition and which requires little or no development effort in order to satisfy a military need (NATO, ARMP-7, 1996).

Off-the-Shelf Item

An item produced and placed in stock by a contractor, or stocked by a distributor, before receiving orders or contracts for its sale. The item may be commercial or produced to military or federal specifications or descriptions (USA DoD, MIL-Std-109C, 1994).

O&M

Operation and Maintenance.

Omission Failure

Failure such that no service is delivered (Laprie, 1992).

On-Delay Timer

A function block delaying a Boolean input signal for a specified duration when changing from zero to one (IEC 1131-1, 1992).

On-Line

Process, to which the Safety Instrumented System is connected, is operating (ISA, S84.01, 1996).

Pertaining to the operation of a functional unit when under the control of a computer (ISO 2382-1, 1993).

A Programmable Controller System is said to be 'on-line' when it is in active control of a machine or process (IEC 1131-1, 1992).

(1) Pertaining to a system or mode of operation in which input data enter the computer directly from the point of origin or output data are transmitted directly to the point where they are used. (2) Pertaining to a device or process that is under the direct control of the central processing unit of a computer (IEEE Std-610.12, 1991).

When equipment or devices operate on-line to an industrial process, they act upon the industrial process and/or are acted upon by this process in the actual time in which the process takes place (IEC 902, 1987).

On-Line Test

An on-line test is a test which is carried out in the on-line operation (VDE 0801, 1990).

Open Loop Control

Control which does not utilise the measurement of the controlled variable (IEC 1131-1, 1992).

Control in which the output variable does not permanently influence the effect of the control itself (IEC 902, 1987).

Open System

A broad term meaning that components of the system conform to internationally agreed, vendor-independent standards allowing easy integration of products from diverse suppliers and smooth interchange of information between them (SSCP 85, 1995).

Operability

The capacity of a system to function. In this Standard, the ability of the system to perform its functions is implied, but in the military context there is often an additional implication of the ability of the operators to use the system effectively (UK MoD, Def Stan 00-58, 1996).

Operable

The state of being able to perform the intended function (USA DoD, MIL-Std-721C, 1992).

Operating Conditions

The loading or demand cyclic operation, or both, of an item between zero and 100% of its related capability(ies) (IEEE Std-500, 1984).

Operating, Not

See Not Operating.

Operating System

Manufacturer's provided software intended for the management of internal Programmable Controller System interdependent functions (IEC 1131-1, 1992).

The machine resident software which enables a computer to function. Without it, application programs could not be loaded or run (Smith & Wood, 1989).

(1) An integrated collection of service routines for supervising the sequencing of programs by a computer. (2) A group of programming systems operating under control of a data-processing monitor program (IEC 902, 1987).

Operating Time

The time during which the system or equipment is turned on and actively performing at least one of its functions (NATO, ARMP-7, 1996).

The period of time that an active item or equipment is functioning effectively (IEEE Std-500, 1984).

The portion of up-time that the system is operating in an acceptable manner (Martz & Waller, 1982).

Operation

A procedural element defining an independent processing activity consisting of the algorithm necessary for the initiation, organisation, and control of phases (ISA, S88.01, 1995).

A defined action, namely, the act of obtaining a result from one or more operands in accordance with a rule that completely specifies the result for any permissible combination of operands (IEC 1131-1, 1992).

The process of using a Target of Evaluation (UK DTI, ITSEC, 1991).

Operation and Maintenance Phase

The period of time in the software life cycle during which a software product is employed in its operational environment, monitored for satisfactory performance, and modified as necessary to correct problems or to respond to changing requirements (IEEE Std-1012, 1987; IEEE Std-729, 1983).

Operation, Proven in

See Proven in Operation.

Operational Availability

The probability that an equipment/system at any instant in the required operating time will operate satisfactorily under stated conditions where the time considered includes operating, corrective and preventive maintenance, administrative delay time and logistic delay time (NATO, ARMP-7, 1996).

The proportion of the defined operational period during which the equipment is available for use without any performance limitations (UK MoD, Def Stan 00-49, 1996).

Operational Environment

The organisational measures, procedures and standards to be used whilst operating a Target of Evaluation (UK DTI, ITSEC, 1991).

Operational Fault

Fault which appears during the system's exploitation (Laprie, 1992).

Operational Mode

One possible operational profile (among two or more) in which program can operate; mode is usually a function of installation or time (Musa et al., 1987).

Operational Profile

The set of all possible input states (input space) with their associated probabilities of occurrence (Musa et al., 1987).

Operational Readiness

The ability of a military unit to respond to its operation plan(s) upon receipt of an operations order. (A function of assigned strength, item availability, status, or supply, training, etc.) (USA DoD, MIL-Std-721C, 1992).

The probability that, at any point in time, the system is either operating or can operate satisfactorily when operated under specified conditions (Martz & Waller, 1982).

Operational Testing

Testing performed for evaluating a system's dependability, with an input profile representative of its operational conditions (Laprie, 1992).

Testing conducted to evaluate a system or component in its operational environment (IEEE Std-610.12, 1991).

Operational Time

The period of time during which an active item performs its intended function, i.e. the accumulated time in service during the period of the event data surveillance (OREDA, 1992).

Operational Version

A software version which has been approved for release to the using community (USA DoD, MIL-Hdbk-347, 1990).

Operator's Workstation

A device that allows the operator to communicate with the PES. It can be used to enter information, to request and display stored data, to actuate various preprogrammed command routines, etc. (CCPS, 1993).

Outage

A state of an item characterised by its inability to perform a required function, for any reason (ESA, ECSS-P-001A, 1997; BSI, BS 4778-3.2, 1991; IEC 50-191, 1990).

Outage Time

(1) Out of service. The average time required to get to the failure, analyse it, obtain spare parts, repair and return the item or equipment to service, including planned delays. (2) Restoration. The average time required to get to the failure, analyse it, obtain spare parts, repair, and return the item or equipment to service, excluding planned delays. (3) Repair. The average time required to analyse the failure, repair, and return the item or equipment to service. This excludes planned delays and waiting for spares or tools (IEEE Std-500, 1984).

Output

(1) The current, voltage, power or driving force delivered by a circuit or device, also called output value. (2) The terminals or other places where current, voltage, power or driving force may be delivered by a circuit or device, also called output terminals. (3) Data which can be written by the application program to an interface between the user program execution function and another functional unit of the Programmable Controller System, as defined in this part of IEC 1131-1. Also called program output (IEC 1131-1, 1992).

(1) Pertaining to data transmitted to an external destination. (2) Pertaining to a device, process, or channel involved in transmitting data to an external destination. (3) To transmit data to an external destination. (4) Loosely, output data (IEEE Std-610.12, 1991).

Output Device

In relation to computers, a device or set of devices designed to convey data out of another device (ASSE, 1988).

Output Module

E/E/PES or subsystem that acts as an interface to external devices and converts output signals into signals that can actuate external devices (ISA, S84.01, 1996).

Output State

Set of output variables for a particular run (Musa et al., 1987).

Output Variable

A variable that exists external to a run and is set by it (Musa et al., 1987).

Overall Quality

The non-quantifiable, qualitative aspects of the safety integrity of the safety-related systems (HSE, 1991).

Overhaul

The effort, usually performed at depot level, when a complete disassembly inspection, rework and re-assembly, of an item is required to restore the item to a 'like new' condition (NATO, ARMP-7, 1996).

Overt

See Detected (IEC 61508-4, 1998).

Overt Fault

Faults that are classified as announced, detected, revealed, etc. (ISA, S84.01, 1996).

P

PADT

See Programming and Debugging Tool.

Parameter

A quantity describing the relation of variables within a given system (IEC 1131-1, 1992; IEC 902, 1987).

(1) A variable that is given a constant value for a specified application. (2) A constant, variable, or expression that is used to pass values between software modules (IEEE Std-610.12, 1991).

A quantity that, when specified, gives specific form to the general form of a model (Musa et al., 1987).

Parameter Estimation

The process of establishing parameter values for a model, based on data (Musa et al., 1987).

Parameter, Planned

See Planned Parameter.

Parameter Prediction

Determination of parameter values from characteristics of the software product and the development process (Musa et al., 1987).

Part

One piece, or two or more pieces joined together which are not normally subject to disassembly without destruction of designed use (USA DoD, MIL-Hdbk-505, 1998; USA DoD, MIL-Std-280A, 1969).

Any hardware item which cannot be disassembled without destroying the capability to perform its required function (ESA, ECSS-P-001A, 1997).

Partitioning, Software

See Software Partitioning.

Parts Count Analysis

A method of predicting item reliability utilising models which need part types and quantity, part quality levels and equipment environment. This prediction method is normally utilised during early design phases and assumes the time to failure of the parts is exponentially distributed, i.e. a constant failure rate (NATO, ARMP-7, 1996).

Pass/Fail Criteria

Decision rules used to determine whether a software item or a software feature passes or fails a test (IEEE Std-829, 1983).

Passivation, Fault

See Fault Passivation.

Passive Fault

Fault which is blocking control actions although the conditions specified in the program are fulfilled (IEC 50-351, 1998).

Passive Safety

The term has long been used in reference to any system that provides safety without human interaction (ASSE, 1988).

Patch

A direct modification of an object module, or a loaded program without assembling or compiling anew from the source program (ISO/IEC DIS 2382-7, 1996).

A modification to an object program, in which one or more of the planned steps of recompiling, reassembling or relinking is bypassed. This does not include identifiers embedded in the software product, e.g. part numbers and checksums (RTCA, DO178B, 1992).

(1) A modification made directly to an object program without reassembling or recompiling from the source program. (2) A modification made to a source program as a last-minute fix or afterthought. (3) Any modification to a source or object program. (4) To perform a modification as in (1), (2), or (3) (IEEE Std-610.12, 1991).

Path

Stream. The order of equipment within a process cell that is used, or is expected to be used, in the production of a specific batch (ISA, S88.01, 1995).

Path Testing

Testing designed to execute all or selected paths through a computer program (IEEE Std-610.12, 1991).

Pathological Coupling

A type of coupling in which one software module affects or depends upon the internal implementation of another (IEEE Std-610.12, 1991).

Pattern Failures

The occurrence of two or more failures of the same part in identical or equivalent applications when the failures are caused by the same basic failure mechanism and the failures occur at a rate which is inconsistent with the parts predicted failure rate (USA DoD, MIL-Hdbk-781A, 1996).

PC

(1) *See* Programmable Controller. (2) Personal Computer.

PC-Program

Synonymous with application program (IEC 1131–1, 1992).

PC-System

Synonymous with Programmable Controller System (IEC 1131–1, 1992).

PCA

Physical Configuration Audit.

PCB

Printed Circuit Board.

PE

See Programmable Electronic.

Penetration Testing

Tests performed by an evaluator on the Target of Evaluation in order to confirm whether or not known vulnerabilities are actually exploitable in practice (UK DTI, ITSEC, 1991).

Perfective Maintenance

Modification of a software product after delivery to improve performance or maintainability (IEEE Std–1219, 1992).

Software maintenance performed to improve the performance, maintainability, or other attributes of a computer program (IEEE Std–610.12, 1991).

Perfective Software Enhancement

Software changes which improve software performance, maintainability or other software attributes. A perfective change does not implement a new system requirement (USA DoD, MIL-Hdbk-347, 1990).

Performability

Performance-related measure of dependability (Laprie, 1992).

Performance

Those generally quantified aspects of an item observed or measured from its operation or function (ESA, ECSS-P-001A, 1997).

The degree to which a system or component accomplishes its designated functions within given constraints, such as speed, accuracy, or memory usage (IEEE Std-610.12, 1991).

Performance Risk

The degree of uncertainty in the development and deployment process that may keep the system from meeting its technical specifications or that may result in the system being unsuitable for its intended use (AFSC/AFLC 00-45, 1988).

Performance Testing

Testing conducted to evaluate the compliance of a system or component with specified performance requirements (IEEE Std-610.12, 1991).

Period, Burn-In

See Burn-In Period.

Period, Constant Failure

See Constant Failure Period.

Period, Early Failure

See Early Failure Period.

Period, Wearout Failure

See Wearout Failure Period.

Peripheral

Catalogued device internal or external to the permanent programmable control system, the function of which is to program, supervise, monitor, test and record the operation of the Programmable Controller System (IEC 1131-1, 1992).

Peripheral Equipment

Any device that is controlled by and can communicate with a particular computer (ISO 2382-1, 1993).

Any unit of equipment, distinct from the central processing unit, which may provide the system with outside communication (IEC 902, 1987).

Peripheral, Permanent

See Permanent Peripheral.

Peripheral, Portable

See Portable Peripheral.

Peripheral, Transportable

See Transportable Peripheral.

Permanent Fault

Fault whose presence is not related to pointwise conditions of the system, either internal or external (Laprie, 1992).

Persistent fault. Solid fault. A fault of an item that persists until an action of corrective maintenance is performed (BSI, BS 4778-3.2, 1991; IEC 50-191, 1990).

Permanent Installation

The portion of the Programmable Controller System which must be present to perform the intended application functions (IEC 1131-1, 1992).

Permanent Peripheral

Peripheral which is part of the permanent installation of the Programmable Controller System (IEC 1131-1, 1992).

Permissive

Condition within a logic sequence that must be satisfied before the sequence is allowed to proceed to the next phase (ISA, S84.01, 1996).

Persistent Fault

See Permanent Fault (BSI, BS 4778-3.2, 1991; IEC 50-191, 1990).

PERT

Program Evaluation Review Technique.

PES

See Programmable Electronic System.

Petri-Net

A method of modelling systems in which activities of one component of the system may occur simultaneously with activities of other components of the system (NATO, ARMP-7, 1996).

An abstract, formal model of information flow, showing static and dynamic properties of a system. A Petri-Net is usually represented as a graph having two types of nodes (called places and transitions) connected by arcs, and markings (called tokens) indicating dynamic properties (IEEE Std–610.12, 1991).

PFD

See Probability of Failure on Demand.

PHA

(1) Process Hazard Analysis. (2) Preliminary Hazard Analysis.

Phase, Concept

See Concept Phase.

Phase, Implementation

See Implementation Phase.

Phase, Installation and Checkout

See Installation and Checkout Phase.

Phase, Manufacturing

See Manufacturing Phase.

Phase, Operation and Maintenance

See Operation and Maintenance Phase.

Phase, Requirements

See Requirements Phase.

Phase, Retirement

See Retirement Phase.

Phase, Test

See Test Phase.

PHI

Process Hazard Identification.

Physical Fault

Fault resulting from adverse physical phenomena (Laprie, 1992).

PID

Proportional-Integral-Derivative.

Plan, Safety

See Safety Plan.

Plan, Test

See Test Plan.

Planned Parameter

Calendar time component parameter relating to resource quantities available and their utilisations, readily determinable from project person responsible (Musa et al., 1987).

PLC

See Programmable Logic Controller.

Portability

The ease with which a system or component can be transferred from one hardware or software environment to another (IEEE Std–610.12, 1991).

Portable Peripheral

Peripheral specifically designed to be carried by hand (IEC 1131-1, 1992).

Post-Mortem Dump

A dump that is produced upon abnormal termination of the execution of a program (ISO/IEC DIS 2382-7, 1996).

A dump that is produced upon abnormal termination of a computer program (IEEE Std-610.12, 1991).

Power Rail

Vertical lines bounding and connected to ladder diagrams on the left and, optionally, on the right (IEC 1131-1, 1992).

Practice

A process, method, or procedure (ASSE, 1988).

Precision

The degree of exactness or discrimination with which a quantity is stated (IEEE Std-610.12, 1991).

A measure of the ability to distinguish between nearly equal values (IEEE Std-729, 1983).

Preconditioning

A method of ageing an item by operating it under specified environmental (normally high temperature or vibration) in order to eliminate early life failures prior to final test or shipment (NATO, ARMP-7, 1996).

Predicted

That which is expected at some future time, postulated on analysis of past experience and tests (USA DoD, MIL-Std-721C, 1992).

Prediction, Error

See Error Prediction.

Prediction Model, Error

See Error Prediction Model.

Prediction, Parameter

See Parameter Prediction.

Predictive Metric

A metric applied during development and used to predict the values of a software quality factor (IEEE Std–1061, 1992).

Preliminary Design

(1) The process of analysing design alternatives and defining the architecture, components, interfaces, and timing and sizing estimates for a system or component. (2) The result of the process in (1) (IEEE Std–610.12, 1991).

Preliminary Design Review

(1) A review conducted to evaluate the progress, technical adequacy, and risk resolution of configuration items; to determine each design's compatibility with the requirements for the configuration item; to evaluate the degree of definition and assess the technical risk associated with the selected manufacturing methods and processes; to establish the existence and compatibility of the physical and functional interfaces among the configuration items and other items of equipment, facilities, software and personnel; and, as applicable, to evaluate the preliminary operational and support documents. (2) A review as in (1) of any hardware or software component (IEEE Std–610.12, 1991).

Preliminary Validation

The tests that are performed on the software specification to ensure that the software specification correctly implements the software requirement (UK MoD, Def Stan 00–55, 1997).

Premature Failure

See Early Life Failure (NATO, ARMP-7, 1996).

Pre-Startup Acceptance Test

Process of confirming performance of the total integrated Safety Instrumented System to assure its conformance to the Safety Requirements Specification and design (ISA, S84.01, 1996).

Prevention Cost

The cost of any action taken to prevent or reduce defects and failures (BSI, BS 4778-2, 1991).

Prevention, Fault

See Fault Prevention.

Preventive Maintenance

Maintenance carried out at predetermined intervals or according to prescribed criteria and intended to reduce the probability of failure or the degradation of the functioning of an item (CENELEC, prEN50126, 1998; BSI, BS 4778-3.2, 1991; IEC 50-191, 1990).

Maintenance performed at predetermined intervals or according to prescribed criteria in order to reduce the probability of failure or the degradation of the functioning of a functional unit (ISO 2382-14, 1997).

Maintenance practice in which equipment is maintained on the basis of a fixed schedule, dictated by manufacturer's recommendation or by accumulated data from operating experience (ISA, S84.01, 1996).

Corrective maintenance aimed at removing faults before they are activated (Laprie, 1992).

The actions performed in an attempt to retain an item in a specified condition by providing systematic inspection, detection, and prevention of incipient failure (USA DoD, MIL-Std-721C, 1992; O'Connor, 1981).

Maintenance performed for the purpose of preventing problems before they occur (IEEE Std-610.12, 1991).

The actions, other than corrective maintenance, carried out for the purpose of keeping an item in a specified condition (Smith, 1981).

Primary Failure

An independent malfunction of equipment under test; a root cause (USA DoD, MIL-Hdbk-781A, 1996).

A failure of an item, not caused either directly or indirectly by a failure or a fault of another item (BSI, BS 4778-3.2, 1991; IEC 50-191, 1990).

Probability

The fraction of occasions on which a specified value or set of values of a quantity occurs, out of all possible values for that quantity (Musa et al., 1987).

Probability Density

Probability per unit variation of random variable (Musa et al., 1987).

Probability Distribution

The set of probabilities corresponding to the values that a random variable can take on (Musa et al., 1987).

Probability of Failure on Demand

A value that indicates the probability of a system failing to respond to a demand. The average probability of a system failing to respond to a demand in a specified time interval is referred to as PFDavg. PFD equals 1 minus Safety Availability (ISA, S84.01, 1996).

The probability of the Instrumented Protective Function failing to respond to a demand. Dimensionless (Korteweg, 1994).

Problem

The inability of a system or component to perform its required functions within specified performance requirements (IEEE Std-1074, 1991).

Procedure

Specified way to perform an activity (ESA, ECSS-P-001A, 1997; ISO 8402, 1994).

A series of activities carried out according to defined rules (UK MoD, Def Stan 00-58, 1996).

A series of activities carried out to accomplish an analysis or assessment (UK MoD, Def Stan 00-56, 1996).

The strategy for carrying out a process (ISA, S88.01, 1995).

(1) A course of action to be taken to perform a given task. (2) A written description of a course of action as in (1). (3) A portion of a computer program that is named and that performs a specific action (IEEE Std-610.12, 1991).

Procedure, Test

See Test Procedure.

Procedure, Unit

See Unit Procedure.

Process

Set of interrelated resources and activities which transform inputs into outputs (ESA, ECSS-P-001A, 1997; ISO 8402, 1994).

A sequence of chemical, physical, or biological activities for the conversion, transport, or storage of material or energy (ISA, S88.01, 1995).

A repeatable set of tasks or activities designed to add value to the specific output of a product/service for a customer (USA DoD, MIL-Std-109C, 1994).

An organised set of activities performed for a given purpose (USA DoD, MIL-Std-498, 1994).

A function that must be performed in the software life cycle. A process is composed of activities (IEEE Std-1074, 1991).

Process Cell

A logical grouping of equipment that includes the equipment required for production of one or more batches. It defines the span of logical control of one set of process equipment within an area (ISA, S88.01, 1995).

Process Computer

Computer directly coupled to a technical process through process interface for data communication in real time (IEC 50-351, 1998).

Process Control

Manipulation of variables influencing the behaviour of a process in order to meet specified objectives (IEC 50-351, 1998).

The control activity that includes the control functions needed to provide sequential, regulatory, and discrete control and to gather and display data (ISA, S88.01, 1995).

The manipulation of variables influencing the behaviour of a process in order to meet specified objectives (IEC 902, 1987).

Process Control System

The complete system to control a process including all equipment, programs and technical regulations (IEC 50-351, 1998).

Process Control System, Basic

See Basic Process Control System.

Process, Development

See Development Process.

Process Hazard Analysis Team

The group of operational, process, instrument/electrical/control, and safety specialists who are responsible for the safety and integrity evaluation of the process from its inception through its implementation and transfer to plant operations, to meet corporate safety guidelines (CCPS, 1993).

Process Measurement

The acquisition of information that establishes the values of process variables (IEC 902, 1987).

Process Metric

A metric used to measure characteristics of the methods, techniques, and tools employed in developing, implementing, and maintaining the software system (IEEE Std-1061, 1992).

Process, Safety

See Safety Process.

Process Safety Audit

An inspection of plant/process unit, drawings, procedures, emergency plan and/or management systems to verify compliance with safety standards (CCPS, 1993).

Process Stage

A part of a process that usually operates independently from other process stages and that usually results in a planned sequence of chemical or physical changes in the material being processed (ISA, S88.01, 1995).

Processing, Error

See Error Processing.

Processing Unit, Main

See Main Processing Unit.

Product

The result of activities or processes (ESA, ECSS-P-001A, 1997; ISO 8402, 1994).

A package of IT software and/or hardware, providing functionality designed for use or incorporation within a multiplicity of systems (UK DTI, ITSEC, 1991).

Product Analysis

The process of evaluating a product by manual or automated means to determine if the product has certain characteristics (IEEE Std–1002, 1987).

Product Baseline

The configuration formally designated at the beginning of production prescribing: (1) all necessary physical and functional characteristics of configuration items; (2) the selected functional characteristics designated for production acceptance testing; and (3) the production acceptance tests (UK MoD, Def Stan 05–57, 1993).

The initially approved documentation describing all of the necessary functional and physical characteristics of the CI, any required joint and combined operations interoperability characteristics of a CI (including a comprehensive summary of the other service(s) and allied interfacing CIs or systems and equipments), and the selected functional and physical characteristics designated for production acceptance testing and tests necessary for support of the CI (USA DoD, MIL-Hdbk-347, 1990; USA DoD, MIL-Std-480B, 1988).

Product Metric

A metric used to measure the characteristics of the documentation and code (IEEE Std-1061, 1992).

Production Permit

A specific written authorisation granted prior to the manufacture of a configuration item, to depart from a particular performance or design requirement of a specification, drawing or other document for a specific number of units or a specific period of time (UK MoD, Def Stan 05-57, 1993).

Profile, Execution

See Execution Profile.

Profile, Life

See Life Profile.

Profile, Life Cycle

See Life Cycle Profile.

Profile, Mission

See Mission Profile.

Profile, Stress

See Stress Profile.

Program

A series of actions proposed in order to achieve a certain result (IEC 1131-1, 1992; IEC 902, 1987).

A set of coded instructions which enable a computer to function. A program may consist of many modules and be written in assembly or high-level language. Note the spelling 'program', whereas 'programme' is used to describe a schedule of tasks (Smith & Wood, 1989).

A set of complete instructions (operators with operands specified) that executes within a single computer and relates to the accomplishment of some major function (Musa et al., 1987).

Program, Application

See Application Program.

Program, Control

See Control Program.

Program, Diagnostic

See Diagnostic Program.

Program Error

Deviation between the function of the program as realised and the program function as intended (Hölscher & Rader, 1986).

Program Memory, User

See User Program Memory.

Program, Sequential

See Sequential Program.

Program, Stable

See Stable Program.

Program, Structured

See Structured Program.

Program, User

See User Program.

Programmable Controller

(For closed loop control.) A microprocessor based controller with a programmable memory for internal storage of user defined instructions (IEC 50–351, 1998).

A digitally operating electronic system, designed for use in an industrial environment, which uses a programmable memory for the internal storage of user-oriented instructions for implementing specific functions such as logic, sequencing, timing, counting and arithmetic, to control through digital or analog inputs and outputs, various types of machines or processes. Both the PC and its associated peripherals are designed so that they can be easily integrated into an industrial control system and easily used in all their intended functions (IEC 1131–1, 1992).

Programmable Controller System

PC-System. A user-built configuration, consisting of a programmable controller and associated peripherals, that is necessary for the intended automated system. It consists of units interconnected by cables or plug-in connections for permanent installation and by cables or other means for portable and transportable peripherals (IEC 1131–1, 1992).

Programmable Device

Any piece of equipment containing one or more components providing a computer architecture with memory facilities (Smith & Wood, 1989).

Programmable Electronic

Based on computer technology which may comprise of hardware, software, and of input and/or output units (IEC 61508-4, 1998).

Programmable Electronic System

System for control, protection or monitoring based on one or more programmable electronic devices, including all elements of the system such as power supplies, sensors and other input devices, data highways and other communication paths, and actuators and other output devices (IEC 61508-4, 1998).

When used in the context of E/E/PES, Programmable Electronic System refers to logic performed by programmable or configurable devices, e.g. Programmable Logic Controller (PLC), Single Loop Digital Controller (SLDC), etc. (ISA, S84.01, 1996).

A system based on one or more central processing units (CPUs), connected to (and including) input devices and/or output devices/final elements (UK MoD, Def Stan 00-58, 1996).

A computer-based system connected to sensors and final control elements for the purpose of control, protection or monitoring (CCPS, 1993).

A system based on a computer connected to sensors and/or actuators on a plant for the purpose of control, protection or monitoring (HSE, 1991).

Programmable Electronic System Controller

A distributed control system controller, PLC system controller, single loop controller, etc. (CCPS, 1993).

Programmable Electronics

Those parts of a PES which are not solely dedicated to a particular sensor or actuator on the plant and in which different functions are performed at different times under the control of software (HSE, 1991).

Programmable Logic Controller

A microprocessor based controller for sequential control, the control logic of which can be changed through a programming device connected to the controller, e.g. programming panel, host computer, hand-held terminal, either directly or remotely through a network (IEC 50–351, 1998).

See Programmable Electronic System (ISA, S84.01, 1996).

A computer, hardened for an industrial environment, for implementing specific functions such as logic, sequencing, timing, counting, and control (CCPS, 1993).

A computer system with real time inputs and outputs. It is provided with a special-purpose problem-oriented language and is increasingly used in the control of plant and processes (Smith & Wood, 1989).

Programming and Debugging Tool

A catalogued peripheral to assist in programming, testing, commissioning and troubleshooting the Programmable Controller System application, program documentation and storage and possibly to be used as MMIs. PADTs are said to be pluggable when they may be plugged or unplugged at any time into their associated interface, without any risk to the operators and the application. In all other cases, PADTs are said to be fixed (IEC 1131–1, 1992).

Programming Error

Error that the programmer introduces into the program during the conversion of the specification into a program (Hölscher & Rader, 1986).

Programming Language

An artificial language for expressing programs (ISO/IEC DIS 2382–7, 1996).

Programming Language, Logic

See Logic Programming Language.

Programming, Structured

See Structured Programming.

Proof Obligation

Mathematical statements arising during a formal design and development process that must be proved in order to verify the design or development step. Also known as verification conditions (UK MoD, Def Stan 00–55, 1997).

Proof of Correctness

A logically sound argument that a program satisfies its requirements (RTCA, DO178B, 1992).

(1) A formal technique used to prove mathematically that a computer program satisfies its specified requirements. (2) A proof that results from applying the technique in (1) (IEEE Std–610.12, 1991).

Proof Test

Periodic test performed to detect failures in a safety-related system so that, if necessary, the system can be restored to an 'as new' condition or as close as practical to this condition (IEC 61508–4, 1998).

Property, Safety

A requirement for a safety feature to be met by the SRS (UK MoD, Def Stan 00–55, 1997).

Protection, Data

See Data Protection.

Protection Layer

Engineered safety features or protective systems or layers that typically involve special process designs, process equipment, administrative procedures, the Basic Process Control System (BPCS), and/or planned responses to protect against an imminent hazard. These responses may be either automated or initiated by human actions (ISA, S84.01, 1996).

Protection layers typically involve special process designs, process equipment, administrative procedures, the basic process control system (BPCS) and/or planned responses to imminent adverse process conditions; and these responses may be either automated or initiated by human actions (CCPS, 1993).

Protection Layer, Independent

See Independent Protection Layer.

Protection System

A system designed to respond to conditions on the plant, which may be hazardous in themselves or if no action were taken could eventually give rise to a hazard, and to generate the correct outputs to mitigate the hazardous consequences or prevent the hazard (HSE, 1991).

Protective Function, Instrumented

See Instrumented Protective Function.

Protective System, Instrumented

See Instrumented Protective System.

Prototype

A preliminary type, form, or instance of a system that serves as a model for later stages or for the final, complete version of the system (IEEE Std-610.12, 1991).

Prototype, Executable

See Executable Prototype.

Proven in Operation

A unit is considered as proven in operation if it was operated essentially over a sufficient period of time in numerous and various applications during which no or very few failures were determined (VDE 0801, 1990).

Proving, Correctness

See Correctness Proving.

Provision

Expression in the content of a normative document, that takes the form of a statement, an instruction, a recommendation or a requirement (ISO/IEC, Guide 2, 1996).

Provisioning

The process of determining and acquiring the quantity and type of support items, spares, repair parts, tools and test equipment necessary to operate and maintain an item for an initially stated period of time (NATO, ARMP-7, 1996).

PSAT

See Pre-Startup Acceptance Test.

PSD

Programmable Safety Device.

Pseudocode

A combination of language constructs from a programming language with those of natural language that is not necessarily computer-processible, but intended to make the design of a program manifest to human readers (ISO/IEC DIS 2382-7, 1996).

PSS

Programmable Safety System.

PSSR

Pre-Startup Safety Review.

Q

QA

See Quality Assurance.

QRA

Quantitative Risk Assessment.

QRA, Chemical Process

The process of hazard identification followed by numerical evaluation of incident consequences and frequencies, and their combination into an overall measure of risk when applied to the chemical process industry (CCPS, 1993).

Qualification

The process of determining whether a system or component is suitable for operational use (IEEE Std–610.12, 1991).

Acknowledgement of the conformity of a model or system to all the requirements of a predetermined set of specifications making it possible to recognise that this model or system is able to assume a specified service (IEC 902, 1987).

Qualification Review, Formal

See Formal Qualification Review.

Qualification, Test

See Test Qualification.

Qualitative Methods

Methods of design and evaluation developed through experience and/or the application of good engineering judgement (ISA, S84.01, 1996).

Methods of design and evaluation developed through experience (CCPS, 1993).

Quality

A user perception of the attributes of a product (CENELEC, ENV50129, 1998).

Totality of characteristics of an entity that bear on its ability to satisfy stated and implied needs (ESA, ECSS-P-001A, 1997; ISO 8402, 1994).

The composite of all the attributes or characteristics, including performance, of an item or product that bear on its ability to satisfy stated or implied needs (USA DoD, MIL-Std-109C, 1994).

The totality of features and characteristics of a product or service that bear on its ability to satisfy stated or implied needs (BSI, BS 4778-2, 1991; BSI, BS 4778-1, 1991).

(1) The degree to which a system, component, or process meets specified requirements. (2) The degree to which a system, component, or process meets customer or user needs or expectations (IEEE Std-610.12, 1991).

Conformance to specification (Smith, 1981).

Quality Assurance

All the planned and systematic activities implemented within the quality system and demonstrated as needed, to provide adequate confidence that an entity will fulfil requirements for quality (ESA, ECSS-P-001A, 1997; ISO 8402, 1994).

A planned and systematic pattern of all actions necessary to provide adequate confidence that management and technical planning and controls are adequate to (1) establish correct technical requirements for design and manufacturing, (2) create products and services that conform to the established technical requirements (USA DoD, MIL-Std-109C, 1994).

(1) A planned and systematic pattern of all actions necessary to provide adequate confidence that an item or product conforms to established technical requirements. (2) A set of activities designed to evaluate the process by which products are developed or manufactured (IEEE Std-610.12, 1991).

All those planned and systematic actions necessary to provide adequate confidence that an item or a facility will satisfy given quality requirements (BSI, BS 4778-2, 1991; IEC 902, 1987).

The total range of activities which attempt to ensure that a finished product, both in design and manufacturing respects, meets the requirement (Smith & Wood, 1989).

Quality Management

That aspect of the overall management function that determines and implements the quality policy (ISO 9000-4, 1993; IEEE Std-1074, 1991; BSI, BS 4778-2, 1991; BSI, BS 4778-1, 1991).

Quality Metric, Software

See Software Quality Metric.

Quality, Overall

See Overall Quality.

Quality Requirement

A requirement that a software attribute be present in software to satisfy a contract, standard, specification, or other formally imposed document (IEEE Std-1061, 1992).

Quantitative Methods

Methods of design and evaluation based on numerical data and mathematical analysis (ISA, S84.01, 1996).

Methods of design and evaluation based on theory and mathematical analysis (CCPS, 1993).

Quantisation

Process by which the range of a variable is divided into a finite number of distinct subranges not necessarily equal, each of which is represented by an assigned value named the quantisation value (IEC 902, 1987).

R

Rail, Power

See Power Rail.

RAM

Reliability, Availability and Maintainability.

RAMS

Reliability, Availability, Maintainability and Safety.

Random

Possessing the property of having more than one value at one time, each occurring with some probability (Musa et al., 1987).

Such that each item has the same probability of being selected as any other (Smith, 1981).

Random Failure

Failures that result from a variety of degradation mechanisms in the hardware. Unlike failures arising from systematic failures, system failure rates arising from random hardware failures can be quantified with reasonable accuracy (UK MoD, Def Stan 00-56, 1996).

Failure whose occurrence is predictable only in a probabilistic or statistical sense (NATO, ARMP-7, 1996; USA DoD, MIL-Std-721C, 1992).

A failure whose occurrence is unpredictable except in a probabilistic or statistical sense (IEEE Std-610.12, 1991).

A failure whose occurrence is statistically dependent on the operating time for the failures of other units (VDE 0801, 1990).

Random Failure Integrity

The degree to which a system is free from hazardous random faults (CENELEC, ENV50129, 1998).

That aspect of the safety integrity relating to random hardware failures (UK MoD, Def Stan 00-56, 1996).

Random Fault

The occurrence of a fault based on probability theory and previous performance (CENELEC, ENV50129, 1998).

Random Hardware Failure

Failure, occurring at a random time, which results from one or more of the possible degradation mechanisms in the hardware (IEC 61508-4, 1998).

Failures which result from a variety of degradation mechanisms in the hardware (HSE, 1991).

Random Process

A set of random variables arranged by some other variable (usually time) (Musa et al., 1987).

Random Testing

See Statistical Testing (Laprie, 1992).

Testing where selection of input states is random with selection probabilities equivalent to operational profile probabilities (Musa et al., 1987).

Random Variable

A variable that possesses the property of randomness (Musa et al., 1987).

Rate, Bit Error

See Bit Error Rate.

Rate, Demand

See Demand Rate.

Rate, Failure

See Failure Rate.

Rate, Fatal Accidents

See Fatal Accidents Rate.

Rate, Hazard

See Hazard Rate.

Rate, Repair

See Repair Rate.

Rate, Utilisation

See Utilisation Rate.

Ratio, Discrimination

See Discrimination Ratio.

Ratio, Failure

See Failure Ratio.

Ratio, Fault Exposure

See Fault Exposure Ratio.

Ratio, Maintenance

See Maintenance Ratio.

Ratio, Success

See Success Ratio.

Ratio, Up-Time

See Up-Time Ratio.

RBD

Reliability Block Diagram.

RCM

Reliability Centred Maintenance.

RCT

Reliability Centred Test.

Reaction Period, Fault

See Fault Reaction Period.

Reaction Time

That element of up-time needed to initiate a mission, measured from the time command is received (NATO, ARMP–7, 1996; USA DoD, MIL-Std–721C, 1992).

Readiness Review, Test

See Test Readiness Review.

Real-Time Function

Function required to be fulfilled within finite time intervals dictated by the environment (Laprie, 1992).

Real-Time Service

Service required to be delivered within finite time intervals dictated by the environment (Laprie, 1992).

Real-Time System

System fulfilling at least one real-time function or delivering at least one real-time service (Laprie, 1992).

Re-assembly

Assembling the items that were removed during disassembly and closing the re-assemble items (USA DoD, MIL-Std–721C, 1992).

Recipe

The necessary set of information that uniquely defines the production requirements for a specific product (ISA, S88.01, 1995).

Recipe, Control

See Control Recipe.

Reconditioning

See Burn-In (USA DoD, MIL-Std-721C, 1992).

Reconfiguration

The modification of the configuration of a functional unit following the detection of a fault or an error for the purpose of preventing a failure or bringing that functional unit back to a state in which it can perform a required function (ISO 2382-14, 1997).

Recover

To establish a previous or new status of all or part of a system, file, database, or other resource, or of the execution of a program so that required functions can be performed (ISO/IEC DIS 2382-7, 1996).

Recovery

Restoration. That event when the item regains the ability to perform a required function (BSI, BS 4778-3.2, 1991; IEC 50-191, 1990).

The restoration of a system, program, database, or other system resource to a state in which it can perform required functions (IEEE Std-610.12, 1991).

Recovery, Backward

See Backward Recovery.

Recovery, Forward

See Forward Recovery.

Recovery Point

Point in time during the execution of a process for which the then current state may subsequently need to be restored (Laprie, 1992).

Recovery Software, Error

See Error Recovery Software.

Recycling

The process by which, after a safety shut-down, a full start-up sequence is automatically repeated (CEN, EN 298, 1993).

Redundancy

Existence of means, in addition to the means which would be sufficient for a functional unit to perform a required function or for data to represent information (IEC 61508-4, 1998; ISO 2382-14, 1997).

In an item, the existence of more than one means for performing a required function (IEC 50-351, 1998).

Use of multiple elements or systems to perform the same function. Redundancy can be implemented by identical elements (identical redundancy) or by diverse elements (diverse redundancy) (ISA, S84.01, 1996).

The employment of two or more devices, each performing the same function, in order to improve reliability (CCPS, 1993).

The performance of the same function by a number of identical but independent means (Jones, 1992).

In an item, the existence of more than one means for performing a given function (IEC 1131-1, 1992).

The existence of more than one means for accomplishing a given function. Each means of accomplishing the function need not necessarily be identical (USA DoD, MIL-Std-721C, 1992; O'Connor, 1981).

In fault tolerance, the presence of auxiliary components in a system to perform the same or similar functions as other elements for the purpose of preventing or recovering from failures (IEEE Std–610.12, 1991).

The availability of more than just the necessary means itself in order to execute the designated tasks (VDE 0801, 1990).

The provision of technical means capable of functioning as required, at a level greater than that necessary for the prescribed function (Hölscher & Rader, 1986).

The provision of more than one means to achieving a function. Active: all items remain operating prior to failure. Stand-by: replicated items do not operate until needed (Smith, 1981).

Redundancy, Active

See Active Redundancy.

Redundancy Computer Configuration

Several computers used in a special configuration, the major number of them are in on-line mode and the others are in stand-by mode, ready to be used if any of the on-line computers malfunction (IEC 902, 1987).

Redundancy, Diverse

See Diverse Redundancy.

Redundancy, Homogeneous

See Homogeneous Redundancy.

Redundancy, Stand-by

See Stand-by Redundancy.

Re-Engineering

The process of examining and altering an existing system to reconstitute it in a new form (USA DoD, MIL-Std-498, 1994).

Refinement, Stepwise

See Stepwise Refinement.

Register

A storage device having a specified storage capacity such as a bit, a byte or a computer word and usually intended for a special purpose (IEC 1131-1, 1992).

Regression Test

Testing to detect faults introduced by the modification (IEEE Std-1219, 1992).

A test case or run that repeats an input state that was previously executed (Musa et al., 1987).

Regression Testing

Selective retesting of a system or component to verify that modifications have not caused unintended effects and that the system or component still complies with its specified requirements (IEEE Std-610.12, 1991).

Regression Verification

Verification performed after a correction, in order to check that the correction has no undesired consequences (Laprie, 1992).

Rejection

A conclusion that a batch, lot or quantity of product or service has not been shown to satisfy the acceptance criteria (BSI, BS 4778-2, 1991).

Rejection Number

The minimum number of defects or defective units in the sample that will cause rejection of the lot represented by the sample (USA DoD, MIL-Std-109C, 1994).

Related Faults

Faults attributed to a common cause (Laprie, 1992).

Relay Ladder Diagram

Synonymous with ladder diagram (IEC 1131-1, 1992).

Release

The designation by the contractor that a document is complete and suitable for use. Release means that the document is subject to the contractor's configuration control procedures (USA DoD, MIL-Std-973, 1992).

A configuration management action whereby a particular version of software is made available for a specific purpose (USA DoD, MIL-Hdbk-347, 1990; USA DoD, MIL-Std-2167A, 1988).

Reliability

The probability that an item can perform a required function under given conditions for a given time interval (t1, t2) (CENELEC, prEN50126, 1998; ESA, ECSS-P-001A, 1997; BSI, BS 4778-3.2, 1991; IEC 50-191, 1990).

The ability of an item to perform a required function under stated conditions for a stated period of time (CENELEC, ENV50129, 1998; SSCP 85, 1995; BSI, BS 4778-1, 1991; Klinger et al., 1990; IEC 902, 1987; O'Connor, 1981).

The ability of a functional unit to perform a required function under given conditions for a given time interval (ISO 2382-14, 1997).

The ability of a system or component to perform its required functions under stated conditions for a specified period of time (UK MoD, Def Stan 00-55, 1997).

Probability that a system can perform a defined function under stated conditions for a given period of time (ISA, S84.01, 1996).

(1) The duration or probability of failure-free performance under stated conditions. (2) The probability that an item can perform its intended function for a specified interval under stated conditions (USA DoD, MIL-Std-109C, 1994; USA DoD, MIL-Std-721C, 1992).

Probability that a component or system will function correctly under stated conditions for a stated period of time (CCPS, 1993).

The probability that an item is able to perform a required function under stated conditions for a stated period of time or for a stated demand (Jones, 1992).

The ability of a functional unit to perform a required function under stated conditions for a stated period of time (IEC 1131-1, 1992; IEEE Std-610.12, 1991).

Dependability with respect to the continuity of service. Measure of continuous correct service delivery. Measure of the time to failure (Laprie, 1992).

That aspect of the safety integrity relating to random hardware failures in a dangerous mode of failure of the safety-related systems (HSE, 1991).

The probability that an item will perform a required function, under stated conditions, for a stated period of time. Since observed reliability is empirical it is defined as the ratio of items which perform their function for the stated period to the total number in the sample (Smith, 1981).

Reliability Block Diagram

A reliability model that provides a pictorial representation of a system reliability performance and shows the logical connection of (functioning) components needed for system success (NATO, ARMP-7, 1996).

Reliability Centred Maintenance

A method for establishing a scheduled (preventive) maintenance programme which will efficiently and effectively achieve the inherent reliability and safety levels of equipment. It is a methodology which can be applied to the development of a preventive maintenance programme and results in improved component reliability and minimised overall programme costs. The intended end result is improved overall equipment safety, availability and economic operation (NATO, ARMP-7, 1996).

Reliability Engineering

That set of design, development, and manufacturing tasks by which reliability is achieved (USA DoD, MIL-Std-785B, 1988).

Reliability Growth

A condition characterised by a progressive improvement of a reliability performance measure of an item with time (CENELEC, prEN50126, 1998; BSI, BS 4778-3.2, 1991; IEC 50-191, 1990).

The improvement in a reliability parameter caused by the successful correction of deficiencies in item design or manufacture (NATO, ARMP-7, 1996; USA DoD, MIL-Std-721C, 1992).

The system's ability to deliver correct service is improved (stochastic increase of the successive times to failure) (Laprie, 1992).

Increase in reliability as a result of continued design modifications resulting from field data feedback (Smith, 1981).

Reliability, Intrinsic

See Intrinsic Reliability.

Reliability Management, Software

See Software Reliability Management.

Reliability, Mission

See Mission Reliability.

Reliability Qualification Test

A test conducted under specified conditions, by, or on behalf of, government, using items representative of the approved production configuration, to determine compliance with specified reliability requirements as a basis for production approval (USA DoD, MIL-Std-785B, 1988).

Reliability, Software

See Software Reliability.

Reliability, Stable

See Stable Reliability.

Relief Valve

A valve designed to release excess pressure within a system without damaging the system itself (ASSE, 1988).

Remote Input/Output Station

RIOS. A manufacturer's catalogued part of a Programmable Controller System including input and/or output interfaces allowed to operate only under the hierarchy of the main processing unit for I/O multiplexing/demultiplexing and data preprocessing/postprocessing. The RIOS is only permitted autonomous operation, e.g. under emergency conditions such as breakdown of the communication link to the MPU or the MPU itself, or when maintenance and troubleshooting operations are to be performed (IEC 1131-1, 1992).

Removal, Fault

See Fault Removal.

Repair

Measures for re-establishing the required state of a system, subsystem or item of equipment (CENELEC, ENV50129, 1998).

Action taken on a non-conforming product so that it will fulfil the intended usage requirements although it may not conform to the originally specified requirements (ESA, ECSS-P-001A, 1997; ISO 8402, 1994).

That part of corrective maintenance in which manual actions are performed on the item (UK MoD, Def Stan 00-49, 1996; BSI, BS 4778-3.2, 1991; IEC 50-191, 1990).

See Corrective Maintenance (USA DoD, MIL-Hdbk-781A, 1996; USA DoD, MIL-Std-721C, 1992).

Repair Coverage

The proportion of faults of an item that can be successfully corrected (BSI, BS 4778-3.2, 1991; IEC 50-191, 1990).

Repair Rate

When used in reliability calculations it is the reciprocal of down-time (Smith, 1981).

Repair Time

The time during which an item is undergoing diagnosis, repair, checkout and alignment. (Must be carefully defined.) (Smith, 1981).

Reparability

The probability that a failed system can be restored to operating condition in a specified interval of active repair time (Martz & Waller, 1982).

Reparable Item

An item which, when capable of being serviced, can normally be economically restored to a serviceable condition through repair procedures performed by a government facility or commercial overhaul facility (USA DoD, MIL-Std-109C, 1994).

Repeatability, Test

See Test Repeatability.

Replaceable Item, Smallest

See Smallest Replaceable Item.

Replaceable, Least

See Least Replaceable.

Replacement in Kind

A replacement which satisfies the design specification (ISA, S84.01, 1996).

Replacement Item

One which is interchangeable with another item, but which differs physically from the original item in that the installation of the replacement item requires operations such as drilling, reaming, cutting, filing, shimming, etc., in addition to the normal application and methods of attachment (USA DoD, MIL-Hdbk-505, 1998; USA DoD, MIL-Std-280A, 1969).

A non-reparable item or one that is not economically feasible to repair (USA DoD, MIL-Std-109C, 1994).

Representation, Design

See Design Representation.

Request, Modification

See Modification Request.

Requirement

That which is called for or is demanded: a condition which must be complied with (ESA, ECSS-P-001A, 1997).

Functional requirements express what the software must do. Non-functional requirements express properties of the final implementation, including accuracy, performance, reliability, user-interface etc (UK MoD, Def Stan 00-55, 1997).

Provision that conveys criteria to be fulfilled (ISO/IEC, Guide 2, 1996).

A detailed statement describing the function and performance of a proposed new system and the domain in which it is to operate (UK MoD, Def Stan 00-58, 1996; UK MoD, Def Stan 00-56, 1996).

(1) A condition or capability needed by a user to solve a problem or achieve an objective. (2) A condition or capability that must be met or possessed by a system or system component to satisfy a contract, standard, specification, or other formally imposed documents. (3) A documented representation of a condition or capability as in (1) or (2) (IEEE Std-610.12, 1991).

Requirement, Functional

See Functional Requirement.

Requirement, Implementation

See Implementation Requirement.

Requirement, Quality

See Quality Requirement.

Requirement, Safety

See Safety Requirement.

Requirement, Software

See Software Requirement.

Requirement Specification

A document that specifies the requirements for a system or component (IEEE Std–610.12, 1991).

Requirements, Derived

See Derived Requirements.

Requirements, High-Level

See High-Level Requirements.

Requirements, Low-Level

See Low-Level Requirements.

Requirements Phase

The period of time in the software life cycle during which the requirements for a software product are defined and documented (IEEE Std–610.12, 1991; IEEE Std–1012, 1987; IEEE Std–729, 1983).

Requirements Review

A process or meeting during which the requirements for a system, hardware item, or software item are presented to project personnel, managers, users, customers, or other interested parties for comment or approval (IEEE Std–610.12, 1991).

Requirements Review, Software

See Software Requirements Review.

Requirements Review, System

See System Requirements Review.

Requirements Specification, Safety

See Safety Requirements Specification.

Requirements Specification, Software

See Software Requirements Specification.

Reset

Action that restores the equipment under control to a predetermined normal enabled or operating state (ISA, S84.01, 1996).

Residual Risk

Risk remaining after protective measures have been taken (IEC 61508-4, 1998).

The risk remaining in a system after completion of the hazard reduction and control process (ESA, ECSS-P-001A, 1997).

The remaining risk after all proposed improvements to the facility under study have been made (Jones, 1992).

Resilience

See Fault Tolerance (ISO 2382-14, 1997).

Resource

Any physically or conceptually identifiable entity whose use and state at any time can be unambiguously determined (ESA, ECSS-P-001A, 1997; BSI, BS 4778-3.2, 1991; IEC 50-191, 1990).

Resource, Shared-User

See Shared-User Resource.

Resource Usage Parameter

Calendar time component parameter that describes resources required for functions such as failure identification and correction (Musa et al., 1987).

Resources, Computer

See Computer Resources.

Restart

To resume the execution of a program using the data recorded at a checkpoint (ISO/IEC DIS 2382-7, 1996).

To cause a computer program to resume execution after a failure, using status and results recorded at a checkpoint (IEEE Std-610.12, 1991).

Restart, Cold

See Cold Restart.

Restart, Hot

See Hot Restart.

Restart, Warm

See Warm Restart.

Restoration

That event when the item regains the ability to perform a required function (UK MoD, Def Stan 00-49, 1996).

See Recovery (BSI, BS 4778-3.2, 1991; IEC 50-191, 1990).

Restoration, Service

See Service Restoration.

Retirement

(1) Permanent removal of a system of component from its operational environment. (2) Removal of support from an operational system or component (IEEE Std–610.12, 1991).

Retirement Phase

The period of time in the software life cycle during which support for a software product is terminated (IEEE Std–610.12, 1991).

Reusability

The degree to which a software module or other work product can be used in more than one computer program or software system (IEEE Std–610.12, 1991).

Reusable Software

Software developed in response to the requirements for one application that can be used, in whole or in part, to satisfy the requirements of another application (USA DoD, MIL-Std–2167A, 1988).

Revealed Failure

A failure whose occurrence is inherently apparent. For Instrumented Protective Functions implemented as non-robust, these failures will cause the Instrumented Protective Function to take protective action (Korteweg, 1994).

A failure, the occurrence of which is obvious by the termination of the ability of the affected item to perform a required function (BSI, BS 4778–3.1, 1991).

Revealed Failure Robust

The intended functionality is not jeopardised with one Revealed Failure (Korteweg, 1994).

Revealed Fault

See Overt Fault (ISA, S84.01, 1996).

Reverse Engineering

The process of extracting software design information from the source code (UK MoD, Def Stan 00–55, 1997).

The process of extracting software system information (including documentation) from source code (IEEE Std–1219, 1992).

Review

Systematic examination of items for the purpose of assessing the results obtained at a given time in the project, by persons not themselves responsible for the project (ESA, ECSS-P–001A, 1997).

A process or meeting during which a work product, or set of work products, is presented to project personnel, managers, users, customers, or other interested parties for comment or approval (IEEE Std–610.12, 1991; IEEE Std–1058.1, 1988).

An evaluation of software element(s) or project status to ascertain discrepancies from planned results and to recommend improvement. This evaluation follows a formal process (IEEE Std–1028, 1988).

Review, Code

See Code Review.

Review, Critical Design

See Critical Design Review.

Review, Design

See Design Review.

Review, Formal Design

See Formal Design Review.

Review, Formal Qualification

See Formal Qualification Review.

Review, Preliminary Design

See Preliminary Design Review.

Review, Requirements

See Requirements Review.

Review, Software Requirements

See Software Requirements Review.

Review, System Design

See System Design Review.

Review, System Requirements

See System Requirements Review.

Review, Test Readiness

See Test Readiness Review.

Rework

A procedure applied to a non-conformance that will completely eliminate it and result in a characteristic that conforms completely to the drawings, specifications, or contract requirements. Rework does not require government approval (USA DoD, MIL-Std-109C, 1994).

RFI

Radio-Frequency Interference.

RFP

Request for Proposal.

Rigorous Argument

An outline of the way in which a proof obligation can be discharged, which presents the main steps but does not supply all the intervening detail (UK MoD, Def Stan 00-55, 1997).

RIOS

See Remote Input/Output Station.

Risk

The combination of the frequency, or probability, and the consequence of a specified hazardous event (CENELEC, ENV50129, 1998).

Combination of the probability of occurrence of harm and the severity of that harm (IEC 61508-4, 1998).

A quantitative measure of the magnitude of a potential loss and the probability of incurring that loss (ESA, ECSS-P-001A, 1997).

The combination of the frequency, or probability, and the consequence of an accident (UK MoD, Def Stan 00-58, 1996; UK MoD, Def Stan 00-56, 1996).

Combination of the frequency, or probability, of occurrence and the consequence of a specified hazardous event (IEC 300-3-9, 1995).

Risk is the hazard level combined with (1) the likelihood of the hazard leading to an accident (sometimes called danger) and (2) hazard exposure or duration (sometimes called latency) (Leveson, 1995).

A measure that combines both the likelihood that a system hazard will cause an accident and the severity of that accident (IEEE Std–1228, 1994).

The combination of Hazard Rate and consequence of a hazardous event (Korteweg, 1994).

A measure of potential human injury, environmental damage, or economic loss in terms of both incident likelihood and the magnitude of the consequent potential injury, damage or loss (CCPS, 1993).

An expression of the possibility/impact of a mishap in terms of hazard severity and hazard probability (USA DoD, MIL-Std–882C, 1993).

The risk which is attached to a particular technical process or condition is described as a summary by a probability statement, in which (1) the expected frequency of the occurrence of an event leads to damage and (2) the extent of the damage which is taken into account and expected when the event takes place (VDE 0801, 1990).

The condition of having outcomes with known probabilities of outcome of occurrence, not certainty of occurrence (AFSC/AFLC 00–45, 1988).

Risk Abatement

The process of reducing the amount of risk to a system (AFSC/AFLC 00–45, 1988).

Risk Acceptance

The acceptance by an individual or organisation of a level or degree of risk which has been identified as the potential consequence of a given course of action (ASSE, 1988).

Risk Analysis

Systematic use of available information to identify hazards and to estimate the risk to individuals or populations, property or the environment (IEC 300-3-9, 1995).

The development of a risk estimate based on engineering evaluation, and mathematical techniques for combining estimates of incident consequences and frequencies (CCPS, 1993).

An imprecise term which infers the quantified calculation of probabilities and risks without taking any judgements about their relevance (Jones, 1992).

Examining the change of outcomes with the modification of the risk drivers. This examination is more involved than risk assessment and should result in the identification of the most critical variables, with insights into desired options for risk handling (AFSC/AFLC 00-45, 1988).

Risk Assessment

Process of making risk estimates and using the results to make decisions (ISA, S84.01, 1996).

Assessment of the system or component to establish that the achieved risk level is lower than or equal to the tolerable risk level (UK MoD, Def Stan 00-56, 1996).

Overall process of risk analysis and risk evaluation (IEC 300-3-9, 1995).

A comprehensive evaluation of the risk and its associated impact (USA DoD, MIL-Std-882C, 1993).

The process of making risk estimates and using the results to make decisions (CCPS, 1993).

The quantitative evaluation of the likelihood of undesired events and the likelihood of harm or damage being caused together with the value judgements made concerning the significance of the results (Jones, 1992).

The process of examining a program and identifying areas of potential risk (AFSC/AFLC 00–45, 1988).

Risk, Consumer's

See Consumer's Risk.

Risk Control

Process of decision-making for managing and/or reducing risk; its implementation, enforcement and re-evaluation from time to time, using the results of risk assessment as input (IEC 300–3–9, 1995).

The process of achieving the desired outcomes by continual application of management techniques to the risk drivers (AFSC/AFLC 00–45, 1988).

Risk, Cost

See Cost Risk.

Risk Drivers

Those variables that cause probabilities of cost, schedule, performance, or support risk to fluctuate significantly (AFSC/AFLC 00–45, 1988).

Risk Estimation

Process used to produce a measure of the level of risks being analysed. Risk estimation consists of the following steps: frequency analysis, consequence analysis and their integration (IEC 300–3–9, 1995).

Risk Evaluation

Process in which judgements are made on the tolerability of the risk on the basis of risk analysis and taking into account factors such as socio-economic and environmental aspects (IEC 300–3–9, 1995).

Risk Handling

The identification of options available to reduce or control selected risk drivers (AFSC/AFLC 00-45, 1988).

Risk, Individual

See Individual Risk.

Risk Level, Tolerable

See Tolerable Risk Level.

Risk Management

Systematic application of management policies, procedures and practices to the tasks of analysing, evaluating and controlling risk (IEC 300-3-9, 1995).

Risk, Residual

See Residual Risk.

Risk, Schedule

See Schedule Risk.

Risk, Societal

See Societal Risk.

Risk, Support

See Support Risk.

Risk, Tolerable

See Tolerable Risk.

R&M

Reliability and Maintenance.

Robbing

See Cannibalise.

Robust, Revealed Failure

See Revealed Failure Robust.

Robust System

System that handles unexpected input states in a way that minimises performance degradation, data corruption, and undesirable output states (Musa et al., 1987).

Robust, Unrevealed Failure

See Unrevealed Failure Robust.

Robustness

The extent to which software can continue to operate correctly despite the introduction of invalid inputs (RTCA, DO178B, 1992).

The degree to which a system or component can function correctly in the presence of invalid inputs or stressful environmental conditions (IEEE Std–610.12, 1991).

This is the ability of a unit, to fulfil the stipulated functions or maintain its operational ability even if the specified marginal conditions are infringed upon. Robustness thus also includes characteristics for which no specific requirements exist (VDE 0801, 1990).

Root Cause

The circumstance that induces or activates the process.

Routine Test

A test to which each individual device is subjected during or after manufacture to ascertain whether it complies with certain criteria (IEC 902, 1987).

Run

(1) In software engineering, a single, usually continuous, execution of a computer program. (2) To execute a computer program (IEEE Std–610.12, 1991).

A subdivision of the execution of a program, usually based on the accomplishment of a user-oriented function (during test, often called a 'test case') (Musa et al., 1987).

Run Time

See Execution Time (ISO/IEC DIS 2382–7, 1996).

Rung

A network in a ladder diagram program with its attached left power rail and optionally attached right power rail (IEC 1131–1, 1992).

S

SA

See Steady Availability.

Safe Failure

Failure which does not have the potential to put the safety-related system in a hazardous or fail-to-function state (IEC 61508–4, 1998).

Safe State

The state of the EUC when safety is achieved (IEC 61508–4, 1998).

A condition which continues to preserve safety (CENELEC, ENV50129, 1998).

State that the equipment under control, or process, shall attain as defined by the Process Hazards Analysis (PHA) (ISA, S84.01, 1996).

Safeguarding

Procedure for actions of the controlling system based on monitoring of the technical process in order to avoid process conditions which would be hazardous to personnel, plant, product or environment (IEC 50-351, 1998).

Safety

Freedom from unacceptable levels of risk (CENELEC, ENV50129, 1998).

Freedom from unacceptable risk (IEC 61508-4, 1998).

State in which the risk of harm (to persons) or damage is limited to an acceptable level (ESA, ECSS-P-001A, 1997; ISO 8402, 1994).

Freedom from unacceptable risk of harm (ISO/IEC, Guide 2, 1996).

The expectation that a system does not, under defined conditions, lead to a state in which human life is endangered (UK MoD, Def Stan 00-58, 1996; UK MoD, Def Stan 00-56, 1996).

The likelihood of a product to maintain throughout its life cycle an acceptable level of risk that may cause an injury to personnel or major damage to the product or its environment (NATO, ARMP-7, 1996).

Safety is freedom from accidents or losses (Leveson, 1995).

Freedom from those conditions that can cause death, injury, occupational illness, or damage to or loss of equipment or property, or damage to the environment (USA DoD, MIL-Std-882C, 1993).

Dependability with respect to the non-occurrence of catastrophic failures. Measure of continuous delivery of either correct service or incorrect service after benign failure. Measure of the time to catastrophic failure (Laprie, 1992).

Safety is a state in which the risk is not greater than the limit risk (VDE 0801, 1990; Hölscher & Rader, 1986).

Safety Analysis

An analysis carried out with the purpose of assessing and examining the safety of the system and its surroundings. A hazard analysis is a necessary element of a safety analysis (UK MoD, Def Stan 00-58, 1996).

Safety Assessment

The process of analysis to determine whether the design authority and the validator have achieved a product that meets the specified safety requirements and to form a judgement as to whether the product is safe for its intended purpose (CENELEC, ENV50129, 1998).

Safety Assessment, Functional

See Functional Safety Assessment.

Safety Assurance

The degree of confidence required in the correctness of a particular process or tool (UK MoD, Def Stan 00-55, 1997).

Safety Audit

A critical examination of all, or part, of a total operating system with relevance to safety (Jones, 1992).

Safety Audit, Functional

See Functional Safety Audit.

Safety Audit, Process

See Process Safety Audit.

Safety Availability

Fraction of time that a safety system is able to perform its designated safety service when the process is operating. In this standard, the average Probability of Failure on Demand (PFDavg) is the preferred term (ISA, S84.01, 1996).

Safety Case

A documented body of evidence that provides a convincing and valid argument that a system is adequately safe for a given application in a given environment (Adelard, ASCAD, 1998).

The documented demonstration that the product complies with the specified safety requirements (CENELEC, ENV50129, 1998).

Safety-Critical

Carries direct responsibility for safety (CENELEC, ENV50129, 1998).

A term used to describe systems, or parts of systems, the failure of which could endanger human life (SSCP 85, 1995).

A term applied to a condition, event, operation, process or item of whose proper recognition, control, performance or tolerance is essential to safe system operation or use; e.g. safety-critical function, safety-critical path, safety-critical component (USA DoD, MIL-Std–882C, 1993).

Safety-Critical Software

Software, including firmware, used to implement a function or component of safety integrity level S4 (UK MoD, Def Stan 00–55, 1997).

Software that falls into one or more of the following categories: (1) Software whose inadvertent response to stimuli, failure to respond when required, response out-of-sequence, or response in combination with other responses can result in an accident. (2) Software that is intended to mitigate the result of an accident. (3) Software that is intended to recover from the result of an accident (IEEE Std–1228, 1994).

Safety Feature

A feature that is required to remove unacceptable safety risks and constrain the other risks from the defence equipment to a tolerable level (includes safety functions and safety properties) (UK MoD, Def Stan 00-55, 1997).

The properties of the system (including its reliability and availability) that are required to remove unacceptable risks and constrain the other risks from the system to a tolerable level (UK MoD, Def Stan 00-56, 1996).

Safety Function

Function to be implemented by an E/E/PE safety-related system, other technology safety-related system or external risk reduction facilities, which is intended to achieve or maintain a safe state for the EUC, in respect of a specific hazardous event (IEC 61508-4, 1998).

A function that the software must perform in order to maintain the safety of the system (UK MoD, Def Stan 00-55, 1997).

Safety, Functional

See Functional Safety.

Safety Instrumented System

System composed of sensors, logic solvers, and final control elements for the purpose of taking the process to a safe state when predetermined conditions are violated. Other terms commonly used include Emergency Shutdown System (ESD, ESS), Safety Shutdown System (SSD), and Safety Interlock System (ISA, S84.01, 1996).

Safety Integrity

Probability of a safety-related system satisfactorily performing the required safety functions under all the stated conditions within a stated period of time (IEC 61508-4, 1998).

The likelihood of a system achieving its required safety features under all the stated conditions within a stated measure of use (UK MoD, Def Stan 00-55, 1997; UK MoD, Def Stan 00-56, 1996).

That characteristic of a safety-related system concerned with its ability to perform its required function, in the desired manner, under all relevant conditions and on the occasions when it is required so to perform (HSE, 1991).

Safety Integrity, Conventional

See Conventional Safety Integrity.

Safety Integrity Criteria

The criteria used as the basis for the safety integrity design and analysis of the safety-related systems (HSE, 1991).

Safety Integrity Level

A number which indicates the required degree of confidence that a system will meet its specified safety features (CENELEC, ENV50129, 1998).

Discrete level (one out of a possible four) for specifying the safety integrity requirements of the safety functions to be allocated to the E/E/PE safety-related systems, where safety integrity level 4 has the highest level of safety integrity and safety integrity level 1 has the lowest (IEC 61508-4, 1998).

An indicator of the required level of safety integrity (UK MoD, Def Stan 00-55, 1997; UK MoD, Def Stan 00-56, 1996).

One of three possible discrete integrity levels (SIL 1, SIL 2, SIL 3) of Safety Instrumented Systems. SILs are defined in terms of probability of failure on demand (PFD) (ISA, S84.01, 1996).

Safety Interlock System

See Safety Instrumented System (ISA, S84.01, 1996).

An instrumented IPL whose purpose is to take the process to a safe state when predetermined conditions are violated (CCPS, 1993).

Safety Interlock System Controller

A controller dedicated to performing SIS functions (CCPS, 1993).

Safety, Intrinsic

See Intrinsic Safety.

Safety Involved

Carries indirect responsibility for safety (CENELEC, ENV50129, 1998).

Safety Life Cycle

Necessary activities involved in the implementation of safety-related systems, occurring during a period of time that starts at the concept phase of a project and finishes when all of the E/E/PE safety-related systems, other technology safety-related systems and external risk reduction facilities are no longer available for use (IEC 61508-4, 1998).

The additional series of activities carried out in conjunction with the system life cycle for safety-related systems (CENELEC, ENV50129, 1998).

Sequence of activities involved in the implementation of the Safety Instrumented System from conception through decommissioning (ISA, S84.01, 1996).

Safety Management

The management structure which ensures that the safety process is properly implemented (CENELEC, ENV50129, 1998).

Safety, Passive

See Passive Safety.

Safety Plan

The implementation details of how the safety requirements of the project will be achieved (CENELEC, ENV50129, 1998).

Safety Process

The series of procedures that are followed to enable all safety requirements of a product to be identified and met (CENELEC, ENV50129, 1998).

Safety Property

A property that is required to hold during execution of the software to achieve safety (UK MoD, Def Stan 00-55, 1997).

Safety-Related

Carries responsibility for safety (CENELEC, ENV50129, 1998).

Safety-Related Software

Software that is used to implement safety functions in a safety-related system (IEC 61508-4, 1998).

Software, including firmware, used to implement a function or component of safety integrity levels S1, S2, S3 or S4 (UK MoD, Def Stan 00-55, 1997).

Safety-Related System

Designated system that both: (1) implements the required safety functions necessary to achieve or maintain a safe state for the EUC; and (2) is intended to achieve, on its own or with other E/E/PE safety-related systems, other technology safety-related systems or external risk reduction facilities, the necessary safety integrity for the required safety functions (IEC 61508-4, 1998).

A system upon which the safety integrity of the plant is to be assured (HSE, 1991).

One or more systems upon which the safety of a plant or process depends (Smith & Wood, 1989).

Safety Requirement

A requirement for a safety feature to be met by the SRS (UK MoD, Def Stan 00–55, 1997).

Safety Requirements Specification

Specification containing all the requirements of the safety functions that have to be performed by the safety-related systems (IEC 61508–4, 1998).

The SRS is a compilation of information found in the PHA report, logic diagram, process technology documents, P&ID, etc., including both functional requirements and safety integrity requirements (CCPS, 1993).

Safety Shutdown System

See Safety Instrumented System (ISA, S84.01, 1996).

Safety, Software

See Software Safety.

Safety Target

A numerical expression of the policy on the tolerability of risks from a system, giving for each identified accident its highest tolerable probability for each group of people who may be harmed by it (UK MoD, Def Stan 00–56, 1996).

Safing

The action of placing a system, or part thereof, in a predetermined safe condition (ESA, ECSS-P-001A, 1997).

SAR

Safety Assessment Report.

SAT

Software Acceptance Test.

SCA

See Sneak Circuit Analysis.

Schedule Risk

The degree of uncertainty associated with the ability of the program to achieve desired milestones (outcomes) on time (AFSC/AFLC 00–45, 1988).

Scheduled Maintenance

Preventive maintenance performed at prescribed points in the item's life (USA DoD, MIL-Std–721C, 1992).

Scheme, Tracking

See Tracking Scheme.

SCM

Software Configuration Management.

SCMP

Software Configuration Management Plan.

SCMPR

Software Configuration Management Plan Review.

SCN

See Specification Change Notice.

SCR

(1) Software Change Request. (2) System Change Request.

Screening

A period of functional and environmental stressing and testing to eliminate manufacturing faults and to reduce the initial failure rate of delivered equipment (UK MoD, Def Stan 00–41, 1993).

A process for inspecting items to remove those that are unsatisfactory or those likely to exhibit early failure. Inspection includes visual examination, physical dimension measurement and functional performance measurement under specified environmental conditions (USA DoD, MIL-Std–721C, 1992).

Screening, Environmental Stress

See Environmental Stress Screening.

Screening Test

A test or a set of tests intended to remove or detect defective items or those likely to exhibit early failures (BSI, BS 4778–3.2, 1991; IEC 50–191, 1990).

SCS

See Safety-Critical Software.

SDD

Software Design Description.

SDP

Software Development Plan.

SDR

See System Design Review.

SECA

Software Error Confinement Area.

Secondary Failure

A failure of an item, caused either directly or indirectly by a failure or a fault of another item (BSI, BS 4778-3.2, 1991; IEC 50-191, 1990).

Security

(Of an electric power system.) Ability of an electric power system to operate in such a way that credible events do not give rise to loss of load, stresses of system components beyond their ratings, bus voltages or system frequency outside tolerances, instability, voltage collapse, or cascading (IEC 50-191, Am. 1, 1999).

Dependability with respect to the prevention of unauthorised access and/or handling of information (Laprie, 1992).

The combination of confidentiality, integrity and availability (UK DTI, ITSEC, 1991).

The technological and administrative safeguards established and applied to a data processing system to protect hardware, software and data from accidental or malicious modifications, destruction or modification (ISO/IEC 2382-8, 1986).

The protection of computer hardware and software from accidental or malicious access, use, modification or disclosure (IEEE Std-729, 1983).

Security Accreditation

The official authorisation which is given to a system to allow it to process classified information in a specific operational environment (SSCP 85, 1995).

Security Certification

The issue of a formal statement, supported by an independent review of the conduct and results of an evaluation, of the extent to which a system meets the security requirement as agreed with the accreditation authority (SSCP 85, 1995).

Security Evaluation

The detailed examination of a computer system to search for vulnerabilities (SSCP 85, 1995).

Security Kernel

A small, self-contained collection of key security-related statements that works as a privileged part of an operating system, specifying and enforcing criteria that must be met for programs and data to be accessed (IEEE Std–610.12, 1991).

SEE

See Software Engineering Environment.

Seeding, Error

See Error Seeding.

Segregation Unit

A physical unit of the system designed such that the immediate effects of the failure of a component that resides in it are restricted to the unit (UK MoD, Def Stan 00–55, 1997; UK MoD, Def Stan 00–56, 1996).

Self-Checking Component

Component comprising error detection mechanisms associated with its functional part (Laprie, 1992).

Semantic Error

An error resulting from a misunderstanding of the relationship of symbols or groups of symbols to their meanings in a given language (IEEE Std–610.12, 1991).

Sensor

A device which, when excited by a physical phenomenon, produces a signal representing the physical phenomenon (IEC 50–351, 1998).

Separation

The use of multiple elements or systems to segregate control from safety functions. Separation can be implemented by identical elements (identical separation) or by diverse elements (diverse separation) (ISA, S84.01, 1996).

The physical and functional isolation of all hardware and software elements. Physical separation is defined as the requirement that the basic process control function (regulatory control - BPCS) and the safety interlock function (SIS) be performed in different logic solvers. Functional separation is achieved through the elimination of common-mode failures in execution of the BPCS and the SIS functions. This may require the separation of BPCS and SIS sensors, final elements, I/O components, the logic solvers, the software operating systems, and the application programs. Some communication may be allowed between separate components and systems as long as no common mode failures can occur (CCPS, 1993).

Sequential

Consisting of components that function one after another (Musa et al., 1987).

Sequential Control System

A control system in which the individual steps are processed in a predetermined order, progression from one sequence to the next being dependent on defined conditions being satisfied. Such a system may be time-dependent, in which the step transition conditions are functions of time only or external-event dependent, where the conditions are functions of input signals only or combinations of these (and perhaps more complex) conditions (IEC 1131-1, 1992).

Sequential Failures

Failures which do not occur within the same predefined time window (Laprie, 1992).

Sequential Function Chart

A graphical representation of a sequential program consisting of interconnected steps (with associated actions), transition conditions and directed links (IEC 1131-1, 1992).

Sequential Program

A plan which describes the actions on a system in a predetermined order and in which some actions depend on the execution of preceding ones or on the fulfilment of certain conditions (IEC 1131-1, 1992).

Serial Data Transmission

A system wherein the bits of a character occur serially in time. Implies only a single transmission channel (IEC 1131-1, 1992).

Service

The result generated by activities at the interface between the supplier and the customer and by supplier internal activities to meet the customer needs (ESA, ECSS-P-001A, 1997; ISO 8402, 1994).

System behaviour as perceived by the system user (Laprie, 1992).

Service, Correct

See Correct Service.

Service, Incorrect

See Incorrect Service.

Service, Real-Time

See Real-Time Service.

Service Restoration

Transition from incorrect to correct service delivery (Laprie, 1992).

Serviceability

The ease with which a system can be repaired (Martz & Waller, 1982).

Servicing

The performance of any act needed to keep an item in operating condition (i.e. lubricating, fuelling, oiling, cleaning, etc.), but not including preventive maintenance of parts or corrective maintenance tasks (USA DoD, MIL-Std-721C, 1992).

Severity

A classification of a failure or undesired event according to the magnitude of its possible consequences (ESA, ECSS-P-001A, 1997).

Grade of the failure consequences upon the system environment (Laprie, 1992).

See Criticality (IEEE Std-610.12, 1991).

Severity, Failure

See Failure Severity.

Severity, Hazard

See Hazard Severity.

SFC

See Sequential Function Chart.

SHA

System Hazard Analysis.

Shared-User Resource

A common resource that can be used by more than one user at a time (ISA, S88.01, 1995).

Shelf Life

The length of time an item can be stored under specified conditions and still meet specified requirements (NATO, ARMP-7, 1996; USA DoD, MIL-Std-721C, 1992).

See Storage Life (USA DoD, MIL-Std-721C, 1992).

Shutdown, Fail-Safe

See Fail-Safe Shutdown.

Shutdown, False

See False Shutdown.

Shutdown System, Emergency

See Emergency Shutdown System.

Side Effect

Actions of a function which affect variables outside the function other than the inputs or outputs (SSCP 83, 1995).

SIFT

Software Implemented Fault Tolerance.

Signal

Physical variable, one or more parameters of which carry information about one or more variables which the signal represents (IEC 902, 1987).

SIL

(1) *See* Software Integrity Level. (2) System Integrity Level.

Simplicity

The degree to which a system or component has a design and implementation that is straightforward and easy to understand (IEEE Std-610.12, 1991).

Simulation

The representation of selected behavioural characteristics of one physical or abstract system by another system (NATO, ARMP-7, 1996).

The use of a data processing system to represent selected behavioural characteristics of a physical or abstract system (ISO 2382-1, 1993).

(1) A model that behaves or operates like a given system when provided a set of controlled inputs. (2) The process of developing or using a model as in (1) (IEEE Std-610.12, 1991).

The representation, for test purposes, of a unit or system by hardware or by software in order to provide some or all inputs or responses (Smith & Wood, 1989).

Simulation Technique, Monte Carlo Computer

See Monte Carlo Computer Simulation Technique.

Simulator

A device, computer program or system used during software verification, that accepts the same inputs and produces the same output as a given system, using object code which is derived from the original source code (RTCA, DO178B, 1992).

A device, computer program, or system that behaves or operates like a given system when provided a set of controlled inputs (IEEE Std-610.12, 1991).

Simultaneous Failures

Failures which occur within some predefined time window (Laprie, 1992).

SIS

(1) *See* Safety Interlock System. (2) Safety Instrumented System.

SLC

Single Loop Controller.

SLCM

Software Life Cycle Model.

Smallest Replaceable Item

The lowest level component at which maintenance will be performed or discrete configuration control enforced (UK MoD, Def Stan 00-56, 1996).

Sneak Circuit Analysis

A Procedure conducted to identify latent paths which cause occurrence of unwanted functions or inhibit desired functions assuming all components are functioning properly (NATO, ARMP-7, 1996; USA DoD, MIL-Std-721C, 1992).

The identification of latent paths which could possibly cause unwanted functions (BSI, BS 4778-3.1, 1991).

Soak Test

A test where a system is submitted to prolonged periods of operation, often at elevated temperature and humidity (Smith & Wood, 1989).

Societal Risk

The relationship between frequency and the number of people suffering from a specified level of harm in a given population from the realisation of specified hazards (Jones, 1992).

Soft Error

See Transient Error (IEEE Std–610.12, 1991).

Soft Fault

Fault for which fault passivation is not undertaken (Laprie, 1992).

Software

Intellectual creation comprising the programs, procedures, data, rules and any associated documentation pertaining to the operation of a data processing system (IEC 61508–4, 1998; ISO 2382–1, 1993).

Computer programs, procedures and associated documentation and data pertaining to the operation of a computer system (UK MoD, Def Stan 00–55, 1997).

Programs, procedures, rules and any associated documentation pertaining to the operation of a computer system (ESA, ECSS-P–001A, 1997).

Computer programs, procedures, rules, associated documentation and data pertaining to the operation of a computer system (UK MoD, Def Stan 05–95, 1995).

Computer programs and computer databases (USA DoD, MIL-Std–498, 1994).

A combination of associated computer instructions and computer data definitions required to enable the computer hardware to perform computational or control functions (USA DoD, MIL-Hdbk-272A, 1993; USA DoD, MIL-Hdbk-347, 1990; USA DoD, MIL-Std-2167A, 1988).

Computer programs, procedures, and possibly associated documentation and data pertaining to the operation of a computer system (IEEE Std-1062, 1992; IEEE Std-610.12, 1991).

(1) Computer programs, routines, programming languages and systems. (2) The collection of related utility, assembly and other programs that are desirable for properly presenting a given machine to a user. (3) Detailed procedures to be followed, whether expressed as programs for a computer or as procedures for an operator or any other person. (4) Documents, including hardware manuals and drawings, computer program listings and diagrams, etc. (5) Items such as those listed above, as contrasted with hardware (IEC 1131-1, 1992).

The whole or part of the program for computing systems whereby the programs permit together with the characteristics of the computing systems (1) the operation of the computing systems, (2) the use of the computing systems in order to solve set tasks or (3) additional operational and application types of the computing system. Related documentation can be considered as a constituent part of the software (VDE 0801, 1990).

The totality or part of the programming constituents of the system, including any documentation which relates to it (Hölscher & Rader, 1986).

Software, Application

See Application Software.

Software Architecture

The structure of the software selected to implement the software requirements (RTCA, DO178B, 1992).

Software Baseline

Software under configuration management, with an agreed and defined build-state or configuration, at a given point in time (UK MoD, Def Stan 05–95, 1995).

Software Change

A modification in source code, object code, executable object code, or its related documentation from its baseline (RTCA, DO178B, 1992).

Software Component

A software item that is placed under configuration control (CSA, CAN/CSA-Q396.1.1, 1989).

Software Component, Computer

See Computer Software Component.

Software Configuration Item

A software item which is under configuration management (UK MoD, Def Stan 05–95, 1995).

Software Correction

A software correction is a software change implemented to correct a fault in the software (USA DoD, MIL-Hdbk–347, 1990).

Software, Critical

See Critical Software.

Software Development

A set of activities that results in software products (USA DoD, MIL-Std–498, 1994).

Software Development Cycle

The period of time that begins with the decision to develop a software product and ends when the software is delivered (IEEE Std-610.12, 1991).

Software Development Process

An organised set of activities performed to translate user needs into software products (USA DoD, MIL-Std-498, 1994).

Software, Diagnostic

See Diagnostic Software.

Software Diversity

A software development technique in which two or more functionally identical variants of a program are developed from the same specification by different programmers or programming teams with the intent of providing error detection, increased reliability, additional documentation, or reduced probability that programming or compiler errors will influence the end results (IEEE Std-610.12, 1991).

Software Documentation, Computer

See Computer Software Documentation.

Software, Embedded

See Embedded Software.

Software Engineering Environment

A host computer system, plus other related equipment and procedures, located in a facility that provides a total support capability of the software of a target computer system (or a set of functionally or physically related target computer systems) (USA DoD, MIL-Hdbk-1467, 1997).

The set of automated tools, firmware devices, and hardware necessary to perform the software engineering effort (USA DoD, MIL-Hdbk-347, 1990; USA DoD, MIL-Std-2167A, 1988).

Software Enhancement

Any software change which is not a software correction (USA DoD, MIL-Hdbk-347, 1990).

Software Enhancement, Adaptive

See Adaptive Software Enhancement.

Software Enhancement, Perfective

See Perfective Software Enhancement.

Software Error

An error in the programming causing a malfunction in use (Smith, 1981).

Software Failure

Deviations of the program functions which are realised in the program and the program functions which are intended (VDE 0801, 1990).

Departure of the external results of program operation from requirements (Musa et al., 1987).

Software, Generic

See Generic Software.

Software Hazard

A software condition that is a prerequisite to an accident (IEEE Std-1228, 1994).

Software Integration

The process of combining code components (RTCA, DO178B, 1992).

Software Item

Any identifiable part of a software product at an intermediate step or at the final step of development (UK MoD, Def Stan 05–95, 1995).

Source code, object code, job control code, control data, or a collection of these items (IEEE Std–610.12, 1991; IEEE Std–829, 1983).

A functionally or logically distinct part of the software (CSA, CAN/CSA-Q396.1.1, 1989).

Software Library

A controlled collection of software and related documentation designed to aid in software development, use, or maintenance (ISO/IEC DIS 2382–7, 1996).

A controlled repository containing a collection of software and related data and documents designed to aid in software development, use or modification (RTCA, DO178B, 1992).

Software Life Cycle

The activities occurring during a period of time that starts when software is conceived and ends when the software is no longer available for use. The software life cycle typically includes a requirements phase, development phase, test phase, integration phase, installation phase and a maintenance phase (CCPS, 1993).

(1) An ordered collection of processes determined by an organisation to be sufficient and adequate to produce a software product. (2) The period of time that begins with decision to produce or modify a software product and ends when the product is retired from service (RTCA, DO178B, 1992).

A project-specific, sequenced mapping of activities (IEEE Std–1074, 1991).

The period of time that begins when a software product is conceived and ends when the software is no longer available for use. The software life cycle typically includes a concept phase, requirements phase, design phase, implementation phase, test phase, installation and checkout phase, operation and maintenance phase, and, sometimes, retirement phase (IEEE Std–610.12, 1991; IEEE Std–1002, 1987).

Software Maintenance

Modification of a software product after delivery to correct faults, to improve performance or other attributes, or to adapt the product to a modified environment (IEEE Std–1219, 1992).

Software Media

A collective name for the physical materials used to hold software and data (UK MoD, Def Stan 05–95, 1995).

Software Module

Construct that consists of procedures and/or data declarations and that can also interact with other such constructs (IEC 61508–4, 1998).

The smallest program unit that is discrete and identifiable with respect to compiling, combining with other units and loading (ESA, ECSS-P–001A, 1997).

Software Partitioning

The process of separating, usually with the express purpose of isolating one or more attributes of the software, to prevent specific interactions and cross-coupling interference (RTCA, DO178B, 1992).

Software Product

See Software (UK MoD, Def Stan 05–95, 1995).

Software or associated information created, modified, or incorporated to satisfy a contract (USA DoD, MIL-Std–498, 1994).

The set of computer programs, and associated documentation and data designated for delivery to a use (RTCA, DO178B, 1992).

(1) The complete set of computer programs, procedures, and possibly associated documentation and data designated for delivery to a user. (2) Any of the individual items in (1) (IEEE Std-610.12, 1991).

Software Quality Metric

A function whose inputs are software data and whose output is a single numerical value that can be interpreted as the degree to which software possesses a given attribute that affects its quality (IEEE Std-1061, 1992).

Software Reliability

The probability that software will not cause the failure of a system for a specified time under specified conditions. The probability is a function of the inputs to and use of the system as well as a function of the existence of faults in the software. The inputs to the system determine whether existing faults, if any, are encountered (IEEE Std-982.1, 1988; IEEE Std-729, 1983).

The probability of failure-free operation of a computer program for a specified time in a specified environment (Musa et al., 1987).

Software Reliability Management

The process of optimising the reliability of software through a program that emphasises software error prevention, fault detection and removal, and the use of measurements to maximise reliability in light of project constraints such as resources (cost), schedule, and performance (IEEE Std-982.1, 1988).

Software Requirement

The system level document defining the requirements which, if met, will ensure that the SRS performs safely and according to operational need (UK MoD, Def Stan 00-55, 1997).

A description of what is to be produced by the software given the inputs and constraints. Software requirements include both high-level requirements and low-level requirements (RTCA, DO178B, 1992).

Software Requirements Review

(1) A review of the requirements specified for one or more software configuration items to evaluate their responsiveness to and interpretation of the system requirements and to determine whether they form a satisfactory basis for proceeding into preliminary design of the configuration items. (2) A review as in (1) for any software component (IEEE Std-610.12, 1991).

Software Requirements Specification

Documentation of the essential requirements (functions, performance, design constraints, and attributes) of the software and its external interfaces (IEEE Std-730, 1989; IEEE Std-1012, 1987).

Software, Reusable

See Reusable Software.

Software Safety

Freedom from software hazards (IEEE Std-1228, 1994).

Software, Safety-Critical

See Safety-Critical Software.

Software, Safety-Related

See Safety-Related Software.

Software Support

The set of activities that takes place to ensure that software installed for operational use continues to perform as intended and fulfil its intended role in system operation. Software support includes software maintenance, aid to users, and related activities (USA DoD, MIL-Std-498, 1994).

The set of all activities that take place to ensure that implemented and fielded software continues to fully support the operational mission of the software (USA DoD, MIL-Hdbk-347, 1990; USA DoD, MIL-Std-2167A, 1988).

Software, Support

See Support Software.

Software System

A system consisting only of software and possibly the computer equipment on which the software operates (USA DoD, MIL-Std-498, 1994).

Software, System

See System Software.

Software Test Environment

A set of automated tools, firmware devices, and hardware necessary to test software (USA DoD, MIL-Hdbk-347, 1990; USA DoD, MIL-Std-2167A, 1988).

Software Test Incident

Any event occurring during the execution of a software test that requires investigation (IEEE Std-610.12, 1991; IEEE Std-1008, 1987).

Software Tool

Software used in the development, testing, analysis, or maintenance of a program or its documentation (ISO/IEC DIS 2382-7, 1996).

A computer program used to help develop, test, analyse, produce, or modify another program or its documentation (UK MoD, Def Stan 05-95, 1995; RTCA, DO178B, 1992).

A computer program used in the development, testing, analysis, or maintenance of a program or its documentation (IEEE Std-610.12, 1991).

Software Unit, Computer

See Computer Software Unit.

Software, Utility

See Utility Software.

Solid Fault

See Hard Fault (Laprie, 1992).

See Permanent Fault (BSI, BS 4778–3.2, 1991; IEC 50–191, 1990).

Solver, Logic

See Logic Solver.

Source Code

Code written in source languages, such as assembly language and/or high-level language, in a machine-readable form for input to an assembler or compiler (RTCA, DO178B, 1992).

Specification

Document stating requirements (ESA, ECSS-P–001A, 1997; ISO 8402, 1994).

The documents that prescribe the requirements of the product or service together with the descriptive means and criteria to assess conformity (UK MoD, Def Stan 00–49, 1996).

A document intended primarily for use in procurement, which clearly and accurately describes the essential and technical requirements for items, materials, or services, including the procedures by which it will be determined that the requirements have been met. Specifications for items and materials may also contain preservation, packaging, packing, and marking requirements (USA DoD, MIL-Std–109C, 1994).

Agreed description of the system's requirements (Laprie, 1992).

The document that prescribes the requirements with which the product or service has to conform (BSI, BS 4778–1, 1991).

A document that specifies, in a complete, precise, verifiable manner, the requirements, design, behaviour, or other characteristics of a system or component, and, often, the procedures for determining whether these provisions have been satisfied (IEEE Std–610.12, 1991).

A detailed description of the parts of a unit and their characteristics with regard to function, performance, quality and safety requirements. This is also a description of the co-operation between these parts and the assumptions made, under which the characteristics are realised (VDE 0801, 1990).

Accurate description of a system in terms of a problem which has to be solved (Hölscher & Rader, 1986).

Specification Change Notice

A document used to propose, transmit, and record changes to a specification (USA DoD, MIL-Std–973, 1992).

Specification, Design

See Design Specification.

Specification Error

Deviation between specified function and intended function (Hölscher & Rader, 1986).

Specification, Formal

See Formal Specification.

Specification, Requirement

See Requirement Specification.

Specification, Safety Requirements

See Safety Requirements Specification.

Specification, Software Requirements

See Software Requirements Specification.

Specification, Technical

See Technical Specification.

Specification, Test

See Test Specification.

Specification, Test Case

See Test Case Specification.

SPMP

Software Project Management Plan.

SPN

Stochastic Petri-Net.

Spurious Trip

Refers to the shutdown of the process for reasons not associated with a problem in the process that the Safety Instrumented System is designed to protect (e.g. the trip resulted due to a hardware fault, software fault, electrical fault, transient, ground plane interference, etc.). Other terms used include nuisance trip and false shutdown (ISA, S84.01, 1996).

SQA

Software Quality Assurance.

SQAP

Software Quality Assurance Plan.

SRR

See Software Requirements Review.

SRS

(1) *See* Software Requirements Specification. (2) *See* Safety-Related System. (3) *See* Safety Requirement Specification.

SSD

See Safety Shutdown System.

SSP

Software Safety Plan.

Stability

A general property of mechanical, electrical, or aerodynamic systems whereby the system returns to a safe state of equilibrium after disturbance, with any consequent oscillation dying away through dissipation of energy (ASSE, 1988).

Stable Program

Program in which code is unchanging, with program neither evolving nor being repaired (Musa et al., 1987).

Stable Reliability

The system's ability to deliver correct service is preserved (stochastic identity of the successive time to failure) (Laprie, 1992).

Stage

Period of execution during which total code and developed code remain constant (Musa et al., 1987).

Stage, Process

See Process Stage.

Standard

A rule or basis of comparison used to provide both guidance in and assessment of the performance of a given activity or the content of a specified data item (RTCA, DO178B, 1992).

Stand-by, Cold

See Cold Stand-by.

Stand-by, Hot

See Hot Stand-by.

Stand-by Mode

That mode of an item during which it is not fully operating but assumed to be completely operable (NATO, ARMP-7, 1996).

Stand-by Operation

A sequence of elementary activities carried out during a quasi-operational state (NATO, ARMP-7, 1996).

Stand-by Redundancy

That redundancy wherein the alternative means of performing the function is inoperative until needed and is switched on upon failure of the primary means of performing the function (USA DoD, MIL-Std-721C, 1992; O'Connor, 1981).

In fault tolerance, the use of redundant elements that are left inoperative until a failure occurs in a primary element (IEEE Std–610.12, 1991).

Starvation

A situation in which an activation of an asynchronous procedure is incapable of proceeding within any predictable interval of time because concurrent asynchronous procedures retain required resources (ISO/IEC DIS 2382–7, 1996).

State

The condition of an equipment entity or of a procedural element at a given time (ISA, S88.01, 1995).

State, Down

See Down-State.

State, Input

See Input State.

State of the Art

Developed stage of technical capability at a given time as regards products, processes and services, based on the relevant consolidated findings of science (ISO/IEC, Guide 2, 1996).

State, Output

See Output State.

State, Safe

See Safe State.

State, System

See System State.

State, Up

See Up-State.

Statement

A programming language construct that represents a set of declarations or a step in a sequence of actions (IEC 1131–1, 1992).

In computer programming, either a meaningful expression that may describe or specify operations or a generalised instruction in automatic coding (IEC 902, 1987).

Statement Coverage

Every statement in the program has been invoked at least once (RTCA, DO178B, 1992).

Statement Testing

Testing designed to execute each statement of a computer program (IEEE Std–610.12, 1991).

Static Analyser

A software tool that helps to reveal certain properties of a program without executing it (RTCA, DO178B, 1992).

Static Analysis

The process of evaluating software without executing the software (UK MoD, Def Stan 00–55, 1997).

Investigating software properties by examining the code without executing the program (SSP 84, 1996; SSCP 83, 1995).

Analysis of computer code by means other than dynamic testing. Static analysis may range from simple visual examination to sophisticated mathematical proving (SSCP 85, 1995).

The process of evaluating a system or component based on its form, structure, content, or documentation (IEEE Std-610.12, 1991).

Static Error

An error that is independent of the time-varying nature of an input (IEEE Std-610.12, 1991).

Static Verification

Verification conducted without exercising the system (Laprie, 1992).

Station, Remote Input/Output

See Remote Input/Output Station.

Statistical Testing

Form of testing where the test patterns are selected according to a defined probability distribution on the input domain (Laprie, 1992).

STE

See Software Test Environment.

Steady Availability

The proportion of time that an item is capable of operating to specification within a large time interval SA=MTBF/(MTBF+MDT) (Smith, 1981).

Stepwise Refinement

A software development technique in which processing steps and data are defined broadly at first, then further defined with increasing detail (ISO/IEC DIS 2382-7, 1996).

A software development technique in which data and processing steps are defined broadly at first and then further defined with increasing detail (IEEE Std–610.12, 1991).

Stopping Failure

System activity, if any, is not any more perceptible to the users, and a constant value service is delivered (Laprie, 1992).

Storage Life

See Shelf Life (USA DoD, MIL-Std–721C, 1992).

Stream

See Path (ISA, S88.01, 1995).

Stress Profile

The degree and number of external influences which a product can withstand whilst performing its required functionality (CENELEC, ENV50129, 1998).

Stress Screening, Environmental

See Environmental Stress Screening.

Stress Test

Marginal Test. Marginal Check. A test in which certain operating conditions are varied about their rated values in order to detect or to locate potential faults (ISO 2382–14, 1997).

Stress Testing

Testing conducted to evaluate a system or component at or beyond the limits of its specified boundaries (IEEE Std–610.12, 1991).

Strong Typing

A feature of a programming language that requires the type of each data item to be declared, precludes the application of operators to inappropriate data types and prevents the interaction of data items of incompatible types (UK MoD, Def Stan 00-55, 1997).

A programming language with strong typing enforces correct types at compile-time, reducing the risk of run-time error (SSP 84, 1996; SSCP 83, 1995).

A feature of some programming languages that requires the type of each data item to be declared, precludes the application of operators to inappropriate data types, and prevents the interaction of data items of incompatible types (IEEE Std-610.12, 1991).

Structural Testing

Form of testing where the testing inputs are selected according to criteria relating to the system's structure (Laprie, 1992).

Testing that takes into account the internal mechanism of a system or component (IEEE Std-610.12, 1991).

Structure

What makes a system what it does (Laprie, 1992).

A specified arrangement or interrelation of parts to form a whole (RTCA, DO178B, 1992).

Structured Design

Any disciplined approach to software design that adheres to specified rules based on principles such as modularity, top-down design, and stepwise refinement of data, of system structures, and of processing steps (ISO/IEC DIS 2382-7, 1996).

A software hierarchical design technique with specified rules governing the hierarchy in terms of design order of modules, independence of modules, and specification of data and interfaces (USA DoD, MIL-Hdbk-272A, 1993).

(1) Any disciplined approach to software design that adheres to specified rules based on principles such as modularity, top-down design, and stepwise refinement of data, system structures, and processing steps. (2) The result of applying the approach in (1) (IEEE Std-610.12, 1991).

Structured Design Method

A disciplined approach that facilitates the partitioning of a software design into a number of manageable stages through decomposition of data and function. Includes object oriented design methods (UK MoD, Def Stan 00-55, 1997).

Structured Program

A program whose control logic topology adheres to strict rules of form, being composed of interactions and nestings of a set of basic planar flowchart constructions (USA DoD, MIL-Hdbk-272A, 1993).

A computer program constructed of a basic set of control structures, each having one entry and one exit (IEEE Std-610.12, 1991).

Structured Programming

(1) A method for constructing programs using only hierarchically arranged constructs each having a single entry point and a single exit point. (2) Any software development technique that includes structured design and that results in the development of structured programs (ISO/IEC DIS 2382-7, 1996).

Any software development technique that includes structured design and results in the development of structured programs (IEEE Std-610.12, 1991).

Structured Text Language

A textual programming language using assignment, subprogram control, selection and iteration statements to represent the application program for a Programmable Controller System (IEC 1131-1, 1992).

Substitute Item

One which possesses such functional and physical characteristics as to be capable of being exchanged for another only under specified conditions or in particular applications and without alteration of the items themselves or of adjoining items (USA DoD, MIL-Hdbk-505, 1998; USA DoD, MIL-Std-280A, 1969).

Subsystem

A portion of a system which fulfils a specialised function (CENELEC, ENV50129, 1998).

A combination of sets, groups, etc. which performs an operational function within a system and is a major subdivision of the system (Examples: data processing subsystem, guidance subsystem) (USA DoD, MIL-Hdbk-505, 1998; USA DoD, MIL-Std-721C, 1992; USA DoD, MIL-Std-280A, 1969).

Set of interdependent elements constituted to achieve a given objective by performing a specified function, but which does not, on its own, satisfy the customer's need (ESA, ECSS-P-001A, 1997).

An element of a system that, in itself may constitute a system (USA DoD, MIL-Std-882C, 1993).

Success Ratio

The probability that an item will perform a required function or that a trial will be successful under stated conditions (IEC 1123, 1991).

Sudden Failure

A failure that could not be anticipated by prior examination or monitoring (BSI, BS 4778-3.2, 1991; IEC 50-191, 1990).

Supervision

The control and monitoring operations of a system including, where appropriate, those operations which ensure reliability and safeguarding (IEC 902, 1987).

Supervisory Control Computer

A computer, with HMI display and input devices, designed for accepting BPCS status inputs and either reporting data to the operator for process manipulation or indirectly regulating the process by manipulating setpoints in the BPCS. Process computer applications realise advanced levels of control, such as economic optimisation, constraint control, model-based control, override control, multivariable control, statistical process control, etc. (CCPS, 1993).

Supplier

Organisation that provides a product to the customer (ESA, ECSS-P-001A, 1997; IEC 300-3-4, 1996; ISO 8402, 1994).

A person or organisation that enters into a contract with the acquirer for the supply of a software product (which may be part of a system) under the terms of a contract (IEEE Std-1061, 1992).

The person, or persons, who produce a product for a customer (IEEE Std-1074, 1991).

Support Equipment

Equipment and computer software required to maintain, test, or operate an item or facility in its intended environment (USA DoD, MIL-Std-973, 1992).

Support Risk

The degree of uncertainty associated with the ability of the support organisation to maintain, change, or enhance software of the fielded system within the planned support concepts and resources (AFSC/AFLC 00-45, 1988).

Support Software

All software used to aid the development, testing, and support of application software. Support software includes all software used to operate, or that is resident in, a Software Engineering Environment (USA DoD, MIL-Hdbk-1467, 1997).

Software or a program that aids in the development, maintenance, or use of other software or provides general application-independent capability (ISO 2382-1, 1993).

Software that aids in the development or maintenance of other software (IEEE Std-610.12, 1991).

Support, Software

See Software Support.

Survey, Hazard

See Hazard Survey.

Survivability

The capability of a system to avoid or withstand a hostile environment without suffering an abortive impairment of its ability to accomplish its designated mission (USA DoD, MIL-Std-973, 1992).

The probability that an item will perform a required function under stated conditions for a stated period of time, without failure (Klinger et al., 1990).

SV&V

Software Verification and Validation.

SVVP

Software Verification and Validation Plan.

SVVPR

Software Verification and Validation Plan Review.

SW

See Software.

Symbolic Execution

Dynamic verification performed with symbolic inputs (Laprie, 1992).

Syntactic Error

A violation of the structural or grammatical rules defined for a language (IEEE Std-610.12, 1991).

Syntax

The rules governing the structure of a language (IEC 1131-1, 1992).

System

A set of subsystems or elements which interact according to a design (CENELEC, ENV50129, 1998).

A set of interrelated elements considered in a defined context as a whole and separated from their environment (IEC 50-351, 1998).

Set of interdependent elements constituted to achieve a given objective by performing a specified function (ESA, ECSS-P-001A, 1997).

A bounded physical entity that achieves in its domain a defined objective through interaction of its parts (UK MoD, Def Stan 00-58, 1996; UK MoD, Def Stan 00-56, 1996).

A composite of equipment and skills, and techniques capable of performing or supporting an operational role, or both. A complete system includes all equipment, related facilities, material, software services, and personnel required for its operation and support to the degree that it can be considered self-sufficient in its intended operational environment (USA DoD, MIL-Std-109C, 1994; USA DoD, MIL-Std-721C, 1992).

A composite, at any level of complexity, of personnel, procedures, materials, tools, equipment, facilities, and software. The elements of this composite entity are used together in the intended operational or support environment to perform a given task or achieve a specific purpose, support, or mission requirement (USA DoD, MIL-Std-882C, 1993).

The total set of subsystems which constitute the system according to the boundary specification (OREDA, 1992).

A set of interlinked units organised to accomplish one or several specific functions (IEEE Std-1219, 1992).

Entity having interacted, interacting, or able to interact with other entities. Set of components bound together in order to interact (Laprie, 1992).

A specific IT installation, with a particular purpose and operational environment (UK DTI, ITSEC, 1991).

Item on the highest level considered in the analysis (IEC 300-3-1, 1991).

A collection of components organised to accomplish a specific function or set of functions (IEEE Std-610.12, 1991).

Set of interdependent elements constituted to achieve a given objective by performing a definite function (IEC 902, 1987).

System, Automated

See Automated System.

System, Automated Verification

See Automated Verification System.

System Design Review

A review conducted to evaluate the manner in which the requirements for a system have been allocated to configuration items, the system engineering process that produced the allocation, the engineering planning for the next phase of the effort, manufacturing considerations, and the planning for production engineering (IEEE Std-610.12, 1991).

System Development Cycle

The period of time that begins with the decision to develop a system and ends when the system is delivered to the end user (IEEE Std-610.12, 1991).

System Effectiveness

The probability that the system can successfully meet an operational demand within a given time period when operated under specified conditions (Martz & Waller, 1982).

A general term covering subjects involving availability, reliability and maintainability (Smith, 1981).

System, Fail-Safe

See Fail-Safe System.

System, Fail-Silent

See Fail-Silent System.

System, Fail-Stop

See Fail-Stop System.

System Hazard

A system condition that is a prerequisite to an accident (IEEE Std-1228, 1994).

System Life Cycle

The series of activities occurring during a period of time that starts when a system is conceived and ends at decommissioning when the system is no longer available for use (CENELEC, ENV50129, 1998).

The period of time that starts when a system is conceived and ends when the system is no longer available for use (IEEE Std-610.12, 1991).

System, Logic

See Logic System.

System, Open

See Open System.

System, Programmable Controller

See Programmable Controller System.

System, Real-Time

See Real-Time System.

System Requirements Review

A review conducted to evaluate the completeness and adequacy of the requirements defined for a system; to evaluate the system engineering process that produced these requirements; to assess the results of system engineering studies; and to evaluate system engineering plans (IEEE Std-610.12, 1991).

System Safety

Freedom from system hazards (IEEE Std-1228, 1994).

The application of engineering and management principles, criteria, and techniques to optimise all aspects of safety within the constraints of operational effectiveness, time, and cost throughout all phases of the system life cycle (USA DoD, MIL-Std-882C, 1993).

An approach to accident prevention which involves the detection of deficiencies in system components which have an accident potential (ASSE, 1988).

System Safety Assessment

An ongoing, systematic, comprehensive evaluation of the proposed system to show that relevant safety-related requirements are satisfied (RTCA, DO178B, 1992).

System, Safety Instrumented

See Safety Instrumented System.

System, Sequential Control

See Sequential Control System.

System Software

Application-independent software that supports the running of application software (ISO 2382–1, 1993).

Software written by the device manufacturer which defines the operation of the device with or without an application program. Typically a collection of subroutines which acts as an interpreter in converting the instructions of the user-entered application program into the machine code by the device hardware (IEC 1131–1, 1992).

Software designed to facilitate the operation and maintenance of a computer system and its associated programs (IEEE Std–610.12, 1991).

System, Software

See Software System.

System State

A condition of being with respect to a set of circumstances (Laprie, 1992).

System Testing

Testing conducted on a complete, integrated system to evaluate the system's compliance with its specified requirements (IEEE Std–610.12, 1991).

The process of testing an integrated hardware and software system to verify that the system meets its specified requirements (IEEE Std–1012, 1987; IEEE Std–729, 1983).

Systematic Event

An event that can be due to faults in the specification, design, construction, operation or maintenance of the system or its components. A systematic event will always cause an event to occur under a particular combination of inputs, or under some particular environmental conditions. A system event that is not caused by a random event is by definition a systematic event. All software failures are systematic events (UK MoD, Def Stan 00–56, 1996).

Systematic Failure

Failure related in a deterministic way to a certain cause, which can only be eliminated by a modification of the design or of the manufacturing process, operational procedures, documentation or other relevant factors (IEC 61508–4, 1998).

Failures due to errors (including mistakes and acts of omissions) in Safety Life Cycle activities that cause the Safety Instrumented System to fail under some particular combination of inputs or under a particular environmental condition. Systematic failures can arise in any Safety Life Cycle step (ISA, S84.01, 1996).

Failures due to errors in design, construction or use of a system which cause it to fail under particular combinations of inputs or under some environmental condition (HSE, 1991).

Systematic Failure Integrity

The degree to which a system is free from non-identified hazardous errors and the causes thereof (CENELEC, ENV50129, 1998).

Systematic Fault

An inherent fault in the specification, design, construction, installation, operation or maintenance of a system, subsystem or component (CENELEC, ENV50129, 1998).

A fault resulting from a systematic failure (BSI, BS 4778-3.2, 1991; IEC 50-191, 1990).

Failure in the realised structure (hardware or software), which are the result of thinking failures ('logical failures') or of failures in execution (including volatile failures) (VDE 0801, 1990).

T

TAAF

Test, Analyse, and Fix.

Target Computer System

Computer equipment, software, and procedures which are physically part of an operational system (USA DoD, MIL-Hdbk-1467, 1997).

Target of Evaluation

An IT system or product which is subjected to security evaluation (UK DTI, ITSEC, 1991).

Task

The basic unit of work from the standpoint of a control program (RTCA, DO178B, 1992).

(1) A sequence of instructions treated as a basic unit of work by the supervisory program of an operating system. (2) In software design, a software component that can operate in parallel with other software components (IEEE Std-610.12, 1991).

TE

See Test Equipment.

Technical Specification

Document that prescribes technical requirements to be fulfilled by a product, process or service (ISO/IEC, Guide 2, 1996).

Technique

The application of a technology – in the present context, a method of carrying out one aspect of a safety or hazard analysis (UK MoD, Def Stan 00–58, 1996).

A method of performing an aspect of an analysis (UK MoD, Def Stan 00–56, 1996).

Temporary Fault

Fault which is present for a limited amount of time (Laprie, 1992).

Terotechnology

An integrated approach to overall optimisation of life cycle costs and resources (Smith, 1981).

Test

A formal process of exercising or putting to trial a system or item by manual or automatic means to identify differences between specified, expected and actual results (ESA, ECSS-P-001A, 1997).

Technical operation that consists of the determination of one or more characteristics of a given product, process or service according to a specified procedure (ISO/IEC, Guide 2, 1996).

(1) An activity in which a system or component is executed under specified conditions, the results are observed or recorded, and an evaluation is made of some aspect of the system or component. (2) To

conduct an activity as in (1). (3) A set of one or more test cases and/or procedures (IEEE Std–610.12, 1991).

The establishments and evaluation of the characteristics of a function unit by means of the stimulated or actual functions being carried out in some defined environment (Hölscher & Rader, 1986).

Test, Acceptance

See Acceptance Test.

Test, Alpha

See Alpha Test.

Test, Automatic

See Automatic Test.

Test, Beta

See Beta Test.

Test, Black Box

See Black Box Test.

Test Case

(1) A set of test inputs, execution conditions, and expected results developed for a particular objective, such as to exercise a particular program path or verify compliance with a specific requirement. (2) Documentation specifying inputs, predicted results, and a set of execution conditions for a test item (IEEE Std–610.12, 1991).

Documentation specifying the details of the test approach for a software feature or combination of software features and identifying the associated tests (IEEE Std–1012, 1987; IEEE Std–829, 1983).

Test Case Generator

A software tool that accepts as input source code, test criteria, specifications, or data structure definitions; uses these inputs to generate test input data; and, sometimes, determines expected results (IEEE Std–610.12, 1991).

Test Case Specification

A document that specifies the test inputs, execution conditions, and predicted results for an item to be tested (IEEE Std–610.12, 1991; IEEE Std–829, 1983).

Test Coverage

The degree to which a given test or set of tests addresses all specified requirements for a given system or component (IEEE Std–610.12, 1991).

Test Criteria

The criteria that a system or component must meet in order to pass a given test (IEEE Std–610.12, 1991).

Test Design

Documentation specifying the details of the test approach for a software feature or combination of software features and identifying the associated tests (IEEE Std–610.12, 1991).

Test Driver

A software module used to invoke a module under test and, often, provide test inputs, control and monitor execution, and report test results (IEEE Std–610.12, 1991).

Test, Endurance

See Endurance Test.

Test Environment, Software

See Software Test Environment.

Test Equipment

A manufacturer's catalogued peripheral equipped with pushbuttons, lamps, keyboards, displays, or equivalent, etc., for troubleshooting of the Programmable Controller System and/or the controlled application (IEC 1131-1, 1992).

Test, Field

See Field Test.

Test, Functional

See Functional Test.

Test Harness

Facility that is capable of simulating (to some useful degree) the operating environment of software under development, by applying test cases to the software and recording the response (IEC 61508-4, 1998).

See Test Driver (IEEE Std-610.12, 1991).

Specially designed hardware and software used to replace parts of the system not yet developed and in order to permit testing of available modules to proceed (Smith & Wood, 1989).

Test Incident, Software

See Software Test Incident.

Test Interval

Time between functional tests (ISA, S84.01, 1996).

Test Item

A software item which is an object of testing (IEEE Std–829, 1983).

Test, Load

See Load Test.

Test Log

A chronological record of all relevant details about the execution of a test (IEEE Std–610.12, 1991; IEEE Std–829, 1983).

Test, Non-Destructive

See Non-Destructive Test.

Test Objective

An identified set of software features to be measured under specified conditions by comparing actual behaviour with the required behaviour described in the software documentation (IEEE Std–610.12, 1991; IEEE Std–1008, 1987).

Test, Off-Line

See Off-Line Test.

Test, On-Line

See On-Line Test.

Test Phase

The period of time in the software life cycle during which the components of a software product are evaluated and integrated, and the software product is evaluated to determine whether or not requirements have been satisfied (IEEE Std–610.12, 1991; IEEE Std–1012, 1987).

Test Plan

A document that describes the technical and management approach to be followed for testing a system or component (IEEE Std–610.12, 1991).

A document describing the scope, approach, resources, and schedule of intended test activities. It identifies test items, the features to be tested, the testing tasks, who will do each task, and any risks requiring contingency planning (IEEE Std–610.12, 1991; IEEE Std–1012, 1987; IEEE Std–829, 1983).

Test, Pre-Startup Acceptance

See Pre-Startup Acceptance Test.

Test Procedure

Detailed instructions for the set-up and execution of a given set of test cases, and instructions for the evaluation of results of executing the test cases (RTCA, DO178B, 1992).

Documentation specifying a sequence of actions for the execution of a test (IEEE Std–610.12, 1991; IEEE Std–1012, 1987; IEEE Std–829, 1983).

(1) Detailed instructions for the set-up, execution, and evaluation of results for a given test case. (2) A document containing a set of associated instructions as in (1) (IEEE Std–610.12, 1991).

Test, Proof

See Proof Test.

Test Qualification

A test conducted under specified conditions, by, or on behalf of the government, using items representative of the production configuration, in order to determine compliance with item design requirements as a basis for production approval (also known as a 'demonstration') (USA DoD, MIL-Std–721C, 1992).

Test Readiness Review

(1) A review conducted to evaluate preliminary test results for one or more configuration items; to verify that the test procedures for each configuration item are complete, comply with test plans and descriptions, and satisfy test requirements; and to verify that a project is prepared to proceed to formal testing of the configuration items. (2) A review as in (1) for any hardware or software component (IEEE Std–610.12, 1991).

Test, Regression

See Regression Test.

Test Repeatability

An attribute of a test, indicating that the same results are produced each time the test is conducted (IEEE Std–610.12, 1991).

Test, Soak

See Soak Test.

Test Specification

Describes the test criteria and the methods to be used in a specific test to assure that the performance and design specifications have been satisfied. The test specification identifies the capabilities or program functions to be tested, and identifies the test environment (IEEE Std–1298, 1992; SA, AS 3563, 1991).

Testing, Statistical

See Statistical Testing.

Test Unit

A set of one or more computer program modules together with associated control data, usage procedures, and operating procedures that satisfy the following conditions: (1) All modules are from a single computer program; (2) At least one of the new or changed modules in the set has not completed the unit test; (3) The set of modules together with its associated data and procedures are the sole object of a testing process (IEEE Std–610.12, 1991; IEEE Std–1008, 1987).

Test, White Box

See White Box Test.

Testability

Testability is a characteristic of design which allows the operational status of an entity, and the location of faulty replaceable components within that entity, to be confidently determined in a timely and cost-effective manner. Operational status can mean operable, partly operable or inoperable (UK MoD, Def Stan 00–49, 1996).

(1) The degree to which a system or component facilitates the establishment of test criteria and the performance of tests to determine whether those criteria have been met. (2) The degree to which a requirement is stated in terms that permit establishment of test criteria and performance of tests to determine whether those criteria have been met (IEEE Std–610.12, 1991).

Testing

An element of inspection that generally denotes the determination by technical means of the properties or elements of supplies, or components thereof, including functional operation, and involves the application of established scientific principles and procedures (USA DoD, MIL-Std–109C, 1994).

Dynamic verification performed with valued inputs (Laprie, 1992).

The process of analysing a software item to detect the differences between existing and required conditions (that is, bugs) and to evaluate the features of the software items (IEEE Std–610.12, 1991; IEEE Std–829, 1983).

The process of operating a system or component under specified conditions, observing or recording the results, and making an evaluation of some aspect of the system or component (IEEE Std–610.12, 1991).

Testing, Acceptance

See Acceptance Testing.

Testing, Back-to-Back

See Back-to-Back Testing.

Testing, Big-Bang

See Big-Bang Testing.

Testing, Black Box

See Black Box Testing.

Testing, Branch

See Branch Testing.

Testing, Component

See Component Testing.

Testing Compression Factor

Ratio of execution time required in operational phase to execution time required in test phase to cover the input space of the program (Musa et al., 1987).

Testing, Deterministic

See Deterministic Testing.

Testing, Development

See Development Testing.

Testing, Dynamic

See Dynamic Testing.

Testing, Fault-Based

See Fault-Based Testing.

Testing, Fault-Finding

See Fault-Finding Testing.

Testing, Formal

See Formal Testing.

Testing, Glass-Box

See Glass-Box Testing.

Testing, Informal

See Informal Testing.

Testing, Integration

See Integration Testing.

Testing, Interface

See Interface Testing.

Testing, Loopback

See Loopback Testing.

Testing, Mutation

See Mutation Testing.

Testing, Operational

See Operational Testing.

Testing, Path

See Path Testing.

Testing, Penetration

See Penetration Testing.

Testing, Performance

See Performance Testing.

Testing, Random

See Random Testing.

Testing, Statement

See Statement Testing.

Testing, Statistical

See Statistical Testing.

Testing, Stress

See Stress Testing.

Testing, Structural

See Structural Testing.

Testing, System

See System Testing.

Testing, Unit

See Unit Testing.

Testing, White Box

See White Box Testing.

Textual Language

A system consisting of a well-defined, usually finite, set of characters; rules for combining characters with one another to form words or other expressions; and a specific assignment of meanings to some of the words or expressions (IEC 1131-1, 1992).

Time, Active

See Active Time.

Time, Active Repair

See Active Repair Time.

Time, Administrative

See Administrative Time.

Time, Alert

See Alert Time.

Time Between Failure, Mean

See Mean Time Between Failure.

Time, Calendar

See Calendar Time.

Time, Checkout

See Checkout Time.

Time, Clock

See Clock Time.

Time, Delay

See Delay Time.

Time, Down

See Down-Time.

Time, Execution

See Execution Time.

Time, Fractional Dead

See Fractional Dead Time.

Time, Free

See Free Time.

Time, Inactive

See Inactive Time.

Time, Logistic

See Logistic Time.

Time, Maintenance

See Maintenance Time.

Time, Mission

See Mission Time.

Time, Modification

See Modification Time.

Time, Not Operating

See Not Operating Time.

Time, Operating

See Operating Time.

Time, Operational

See Operational Time.

Time, Outage

See Outage Time.

Time, Reaction

See Reaction Time.

Time, Repair

See Repair Time.

Time To Failure, Mean

See Mean Time To Failure.

Time To Repair, Mean

See Mean Time To Repair.

Time, Turn Around

See Turn Around Time.

Time, Up

See Up-Time.

Timer, Off-Delay

See Off-Delay Timer.

Timer, On-Delay

See On-Delay Timer.

Timer, Watchdog

See Watchdog Timer.

Timing Failure

Failure such that the timing of service delivery does not comply with the specification (Laprie, 1992).

TOE

Target of Evaluation.

Tolerable Risk

Risk which is accepted in a given context based on the current values of society (IEC 61508–4, 1998).

Tolerable Risk Level

The maximum level of risk of a particular technical process or condition that is regarded as tolerable in the circumstances in question (UK MoD, Def Stan 00–56, 1996).

Tolerance, Error

See Error Tolerance.

Tolerance, Fault

See Fault Tolerance.

Tool

Software packages for the development, management, integration and validation of software and for all project support activities (SSCP 85, 1995).

A product used in the construction and/or documentation of a Target of Evaluation (UK DTI, ITSEC, 1991).

Tool, Programming and Debugging

See Programming and Debugging Tool.

Tool Qualification

The process necessary to obtain certification credit for a software tool within the context of a specific airborne system (RTCA, DO178B, 1992).

Tool, Software

See Software Tool.

Top Event

The selected outcome whose possible causes are analysed in a fault tree (Jones, 1992).

Total Correctness

In proof of correctness, a designation indicating that a program's output assertions follow logically from its input assertions and processing steps, and that, in addition, the program terminates under all specified input conditions (IEEE Std-610.12, 1991).

Total Failures

Number of failures that will occur in infinite execution time (Musa et al., 1987).

Traceability

Ability to trace the history, application or location of an entity by means of recorded identifications (ISO 8402, 1994).

The evidence of an association between elements (such as between process outputs, an output and its originating process, a requirements specification and its implementation, etc.) (RTCA, DO178B, 1992).

The ability to trace the history, application or location of an activity, or similar items or activities, by means of recorded identification (BSI, BS 4778-1, 1991).

(1) The degree to which a relationship can be established between two or more products of the development process, especially products having a predecessor-successor or master-subordinate relationship to one another. (2) The degree to which each element in a software development product establishes its reason for existing (IEEE Std-610.12, 1991).

Tracking Scheme

A scheme for documenting the inheritance of integrity requirements by the system components during the design (UK MoD, Def Stan 00-56, 1996).

Train

Line. A collection of one or more units and associated lower level equipment groupings that has the ability to be used to make a batch of material (ISA, S88.01, 1995).

Transient Error

An error that occurs once, or at unpredictable intervals (IEEE Std-610.12, 1991).

Transient Fault

Temporary physical external fault (Laprie, 1992).

Transmission, Serial Data

See Serial Data Transmission.

Transportable Peripheral

Peripheral which is not a permanent part of the installation and which is connectable to the system by means of cables or cords with connectors or other means to establish communication with the system (IEC 1131-1, 1992).

Treatment, Fault

See Fault Treatment.

Trip, De-Energise To

See De-Energise To Trip.

Trip, Energise To

See Energise To Trip.

Trip, Nuisance

See Nuisance Trip.

Trip, Spurious

See Spurious Trip.

Tripping, Cascade

See Cascade Tripping.

TRR

See Test Readiness Review.

Trustability

System's ability to provide users with information about service correctness (Laprie, 1992).

Turn Around Time

That element of maintenance time needed to replenish consumables and check out an item for recommitment (USA DoD, MIL-Std-721C, 1992).

Two-Hand Controls

Tripping devices for a machine which require the simultaneous application of both hands to operate the control, thus assuring that the operator's hands are kept out of the point-of-operation area while the machine is being used (ASSE, 1988).

Type Approval

Approval based on type testing (ISO/IEC, Guide 2, 1996).

Type Test

A test of one or more devices made to a certain design to show that the design meets certain specifications (IEC 902, 1987).

Type Testing

Conformity testing on the basis of one or more specimens of a product representative of the production (ISO/IEC, Guide 2, 1996).

Typing, Strong

See Strong Typing.

U

UA

User Approved.

UAS

User Approved Safety.

Unconditional Jump

In the execution of computer program, a departure from the implied or declared order in which instructions are being executed (IEC 1131-1, 1992).

Undetected

Unrevealed. Covert. In relation to hardware, undetected by the diagnostic tests, proof tests, operator intervention (for example physical inspection and manual tests), or through normal operation (IEC 61508-4, 1998).

Unit

A collection of associated control modules and/or equipment modules and other process equipment in which one or more major processing activities can be conducted (ISA, S88.01, 1995).

(1) A separately testable element specified in the design of a computer software component. (2) A logically separable part of a computer program. (3) A software component that is not subdivided into other components (IEEE Std-610.12, 1991).

See Test Unit (IEEE Std-1008, 1987).

Unit, Functional

See Functional Unit.

Unit Procedure

A strategy for carrying out a contiguous process within a unit. It consists of contiguous operations and the algorithm necessary for the initiation, organisation, and control of those operations (ISA, S88.01, 1995).

Unit, Segregation

See Segregation Unit.

Unit, Test

See Test Unit.

Unit Testing

Testing of individual hardware or software units or groups of related units (IEEE Std–610.12, 1991).

Units, Life

See Life Units.

Unrevealed

See Undetected (IEC 61508–4, 1998).

Unrevealed Failure

A failure which is dormant in the Instrumented Protective Function and can only be found as a result of a need for the system to take action or through testing (Korteweg, 1994).

Any failure whose occurrence can only be detected by failure to function on demand or by inspection or test (BSI, BS 4778–3.1, 1991).

Unrevealed Failure Robust

The intended functionality is not jeopardised with one Unrevealed Failure (Korteweg, 1994).

Unrevealed Fault

See Covert Fault (ISA, S84.01, 1996).

Unscheduled Maintenance

Corrective maintenance required by item conditions (USA DoD, MIL-Std-721C, 1992).

The maintenance carried out, not in accordance with an established time schedule, but after reception of an indication regarding the state of an item (BSI, BS 4778-3.2, 1991; IEC 50-191, 1990).

Up-State

A state of an item characterised by the fact that it can perform a required function, assuming that the external resources, if required, are provided (BSI, BS 4778-3.2, 1991; IEC 50-191, 1990).

UPS

Uninterruptable Power Supply.

Up-Time

That period of time during which an item is in a condition to perform a required function (NATO, ARMP-7, 1996).

That element of active time during which an item is in condition to perform its required functions (UK MoD, Def Stan 00-49, 1996; USA DoD, MIL-Std-721C, 1992).

The period of time during which a system or component is operational and in service; that is, the sum of busy time and idle time (IEEE Std-610.12, 1991).

The total time during which the system is in acceptable operating condition (Martz & Waller, 1982).

Up-Time Ratio

A composite measure of operational availability and dependability that includes the combined effects of item design, installation, quality, environment, operation, maintenance, repair and logistic support: the quotient of up-time divided by up-time plus down-time (USA DoD, MIL-Std-721C, 1992).

Useful Life

The number of life units from manufacture to when the item has an unreparable failure or unacceptable failure rate (USA DoD, MIL-Std-721C, 1992).

User

The individual(s) who, or organisation(s) or service(s) that operate(s) or use(s) the product (UK MoD, Def Stan 00-49, 1996).

Another system (physical, human) interacting with the considered system (Laprie, 1992).

The person, or persons, who operate or interact directly with the system (IEEE Std-1219, 1992; IEEE Std-1074, 1991).

User Approved

Hardware, software, procedures, etc. that the user has evaluated and determined to be acceptable for the application (ISA, S84.01, 1996).

User Interface

An interface that enables information to be passed between a human user and hardware or software components of a computer system (IEEE Std-610.12, 1991).

User Interface, Graphical

See Graphical User Interface.

User Program

Synonymous with Application Program (IEC 1131-1, 1992).

User Program Memory

The portion of the PC memory reserved for the storage of application programs (IEC 1131-1, 1992).

Utilisation

Proportion of time a resource is used (Musa et al., 1987).

Utilisation, Computer

See Computer Utilisation.

Utilisation, Memory

See Memory Utilisation.

Utilisation Rate

The planned or actual number of life units expended, or missions attempted during a stated interval of calendar time (USA DoD, MIL-Std-721C, 1992).

Utility Software

Software tools for the creation, maintenance, and documentation of application programs. These software tools are not required for the operation of the Safety Instrumented System (ISA, S84.01, 1996).

V

Validated Metric

A metric whose values have been statistically associated with corresponding quality factor values (IEEE Std-1061, 1992).

Validation

The activity of demonstration, by test and analysis, that the product meets in all respects its specified requirements (CENELEC, ENV50129, 1998).

Confirmation by examination and provision of objective evidence that the particular requirements for a specific intended use are fulfilled (IEC 61508-4, 1998; ESA, ECSS-P-001A, 1997; ISO 8402, 1994).

The process of evaluating software at the end of the software development process to ensure compliance with software requirements (UK MoD, Def Stan 00-55, 1997; IEEE Std-1012, 1987; IEEE Std-729, 1983).

The process of checking that a system meets its requirements. In particular compiler validation is the process of checking that a compiler conforms to an appropriate standard (SSP 84, 1996).

The process of determining that software complies with specified requirements (UK MoD, Def Stan 05-95, 1995).

The process of evaluating software during the development process to ensure compliance with both hardware and software requirements (SSCP 85, 1995).

The process of determining that the requirements are the right requirements and that they are complete. The system process may use software requirements and derived requirements in system validation (RTCA, DO178B, 1992).

Methods and techniques intended to enable confidence to be reached in a system's ability to deliver a service complying with the specifications. Fault removal and fault forecasting (Laprie, 1992).

The process of evaluating a system or component during or at the end of the development process to determine whether it satisfies specified requirements (IEEE Std-610.12, 1991).

The process of evaluating software to determine compliance with specified requirements (USA DoD, MIL-Hdbk-347, 1990; USA DoD, MIL-Std-2167A, 1988).

Validation, Independent Verification and

See Independent Verification and Validation.

Validation, Preliminary

See Preliminary Validation.

Validation, Verification and

See Verification and Validation.

Validity

Validity has been achieved when the specification meets the objectives which have been set (Hölscher & Rader, 1986).

Value Failure

Failure such that the value of the delivered service does not comply with the specification (Laprie, 1992).

Valve, Relief

See Relief Valve.

Variable

A language object that may take different values, one at a time (IEC 1131-1, 1992).

A quantity or condition whose value is subject to change and can usually be measured (IEC 1131-1, 1992; IEC 902, 1987).

Variable, Input

See Input Variable.

Variable, Output

See Output Variable.

Variable, Random

See Random Variable.

Variant

In fault tolerance, a version of a program resulting from the application of software diversity (IEEE Std–610.12, 1991).

VDU

Video Display Unit.

Verifiability, Design for

See Design for Verifiability.

Verification

Confirmation by examination and provision of objective evidence that the requirements have been fulfilled (IEC 61508–4, 1998).

Confirmation by examination and provision of objective evidence that specified requirements have been fulfilled (ESA, ECSS-P–001A, 1997; IEC 300–3–4, 1996; ISO 8402, 1994).

The process of evaluating a system or component to determine whether the products of a given development phase satisfy the conditions imposed at the start of that phase (UK MoD, Def Stan 00–55, 1997).

Process of confirming for certain steps of the safety life cycle that the objectives are met (ISA, S84.01, 1996).

The process of determining whether or not the product of each phase of the computer system development process is an accurate realisation of the one before (SSP 84, 1996).

The process of determining that the products of a given software design and development phase are correct and consistent with respect to the products and standards provided as input to that phase (UK MoD, Def Stan 05–95, 1995).

The process of determining whether or not the products of a given phase of software development fulfil the requirements established during the previous phase (SSCP 85, 1995).

The art of reviewing, inspecting, testing, checking, auditing or otherwise establishing and documenting whether items, processes, services, or documents conform to specified requirements (USA DoD, MIL-Std-109C, 1994).

The process of determining whether a system adheres to properties (the verification conditions) which can be (1) general, independent of the specification, or (2) specific deduced from the specification (Laprie, 1992).

(1) The process of evaluating a system or component to determine whether the products of a given development phase satisfy the conditions imposed at the start of that phase. (2) Formal proof of program correctness (IEEE Std-610.12, 1991).

The process of evaluating the products of a given software development activity to determine correctness and consistency with respect to the products and standards provided as input to that activity (USA DoD, MIL-Hdbk-347, 1990; USA DoD, MIL-Std-2167A, 1988).

The process of determining whether or not the products of a given phase of the software development cycle fulfil the requirements established during the previous phase (IEEE Std-1012, 1987; IEEE Std-729, 1983).

The proof test of the fact that the operations of a unit match the specification (Hölscher & Rader, 1986).

Verification and Validation

The process of determining whether the requirements for a system or component are complete and correct, the products of each development phase fulfil the requirements or conditions imposed by the previous phase, and the final system or component complies with specified requirements (IEEE Std-610.12, 1991; IEEE Std-1002, 1987).

Verification and Validation, Independent

See Independent Verification and Validation.

Verification, Dynamic

See Dynamic Verification.

Verification, Regression

See Regression Verification.

Verification, Static

See Static Verification.

Verification System, Automated

See Automated Verification System.

Version

(1) An initial release or re-release of a computer software configuration item, associated with a complete compilation or recompilation of the computer software configuration item. (2) An initial release or complete re-release of a document, as opposed to a revision resulting from issuing change pages to a previous release (IEEE Std–610.12, 1991).

An identified and documented body of software. Modifications to a version of software (resulting in a new version) require configuration management actions by either the contractor, the contracting agency, or both (USA DoD, MIL-Hdbk–347, 1990; USA DoD, MIL-Std–2167A, 1988).

Version, Operational

See Operational Version.

Volatile Lock-Out

The safety shut-down condition of the system, such that a restart can only be accomplished by either the manual reset of the system, or an interruption of the main power and its subsequent restoration (CEN, EN 298, 1993).

Volatile Memory

A storage medium that loses information when power is removed from the system (USA DoD, MIL-Hdbk-272A, 1993).

Voting System

Redundant system (e.g. m out of n, one out of two [1oo2] to trip, two out of three [2oo3], etc.) which requires at least 'm' of the 'n' channels to be in agreement before the Safety Instrumented System can take an action (ISA, S84.01, 1996).

Vulnerability

A security weakness in a TOE (for example, due to failures in analysis, design, implementation or operation) (UK DTI, ITSEC, 1991).

V&V

See Verification and Validation.

W

Waiver

Written authorisation to use or release a product which does not conform to the specified requirements (ESA, ECSS-P-001A, 1997; ISO 8402, 1994).

See Concession (UK MoD, Def Stan 05-57, 1993; BSI, BS 4778-1, 1991).

A written authorisation to accept an item, which during manufacture, or after having been submitted for government inspection or acceptance, is found to depart from specified requirements, but nevertheless is considered suitable for use 'as is' or after repair by an approved method (USA DoD, MIL-Std-973, 1992).

Walk-Through

A static analysis technique in which a designer or programmer leads members of the development team and other interested parties through a segment of documentation or code, and the participants ask questions and make comments about possible errors, violation of development standards, and other problems (IEEE Std-610.12, 1991).

Warm Restart

Restart of a system after Power Failure with a user programmed predetermined set of dynamic data and a system predetermined application program context. A warm restart is identified by a status flag or equivalent means made available to the application indicating that the power failure shutdown of the system was detected in the 'run' mode (IEC 1131-1, 1992).

Warning

Any of various means used to advise of or call attention to a danger which should be heeded if one is to avoid possible damage or injury (ASSE, 1988).

Advisory in a software user document that performing some action will lead to serious or dangerous consequences (IEEE Std-1063, 1988).

Watchdog

(1) A manufacturer-provided means performing specific actions if the integrity of the Programmable Controller System is violated. (2) A manufacturer-provided means which independently monitors the duration of internal hardware functions, and/or application program functions, and/or operating system software functions, and which will

cause specific actions to be performed if not periodically reset at a predetermined interval (IEC 1131-1, 1992).

A part of a system (may be hardware or software) which monitors the state of the processor and signals when tasks are not completed within a prescribed time (Smith & Wood, 1989).

Watchdog Timer

A timer which monitors the state of a signal or a functional unit for inactivity or delay of response beyond a specified duration (ISO 2382-14, 1997).

A timer implemented to prevent the system from looping endlessly, providing inaccurate communications, or becoming idle because of program errors or equipment faults (CCPS, 1993).

A timer set by the program to prevent the system from looping endlessly or becoming idle because of program errors or equipment faults (IEC 902, 1987).

WDT

See Watchdog Timer.

Weak Typing

A programming language with weak typing does not check whether types are correct at compile-time, with a consequent risk of run-time faults (SSP 84, 1996; SSCP 83, 1995).

Wearout

The process which results in an increase of the failure rate or probability of failure with increasing number of life units (USA DoD, MIL-Std-721C, 1992).

Wearout Failure

A failure whose probability of occurrence increases with the passage of time (NATO, ARMP-7, 1996; BSI, BS 4778-3.2, 1991; IEC 50-191, 1990).

Wearout Failure Period

The period in the life cycle of a system or component during which hardware failures occur at an increasing rate due to detoriation (IEEE Std-610.12, 1991).

What-If Analysis

A systematic approach to hazard analysis that asks 'What if this, that, or some other event might occur?' in order to identify every conceivable cause of harm that could arise in every aspect of a given process or activity. Each identified cause is then quantified as to the probability of its occurring, and measures to control or counteract it are devised (ASSE, 1988).

White Box

See Glass Box (IEEE Std-610.12, 1991).

White Box Test

See Structural Testing (IEEE Std-610.12, 1991).

White Box Testing

Testing techniques which use knowledge of the internal structure of the item under test. Techniques based on knowledge of the control flow are most common (UK MoD, Def Stan 00-55, 1997).

Workstation, Operator's

See Operator's Workstation.

References

AFSC/AFLC 00-45, 1988
Department of the Air Force, Headquarters Air Force Systems Command, Andrews Air Force Base DC 20334-5000 Headquarters Air Force Logistics Command, Wright-Patterson Air Force Base OH 45433-5001, *Software risk abatement*, AFSC/AFLC Pamphlet 00-45, 30 September 1988.

Adelard, ASCAD, 1998
Adelard, *Adelard safety case development manual*, 1998.

ASSE, 1988
American Society of Safety Engineers, *The dictionary of terms used in the safety profession*, 3rd edition, 1988.

BSI, BS 4778-1, 1991
British Standards Institution, BS 4778, Part 1, *Quality vocabulary – International terms*, 1991.

BSI, BS 4778-2, 1991
British Standards Institution, BS 4778, Part 2, *Quality concepts and related definitions*, 1991.

BSI, BS 4778-3.1, 1991
British Standards Institution, BS 4778, Part 3, *Availability, reliability and maintainability terms*; Section 3.1: 'Guide to concepts and related definitions', 1991.

BSI, BS 4778-3.2, 1991
British Standards Institution, BS 4778, Part 3, *Availability, reliability and maintainability terms*; Section 3.2: 'Glossary of international terms', 1991 (equal to IEC 50-191, 1990).

CSA, CAN/CSA-Q396.1.1, 1989
CSA International, CAN/CSA-Q396.1.1-89, *Quality Assurance Program for the development of software used in critical applications*, 1989.

CCPS, 1993
Center for Chemical Process Safety of the American Institute of Chemical Engineers, *Guidelines for safe automation of chemical processes*, New York, 1993. Copyright (1992) by the American Institute of Chemical Engineers and reproduced by permission of Center for Chemical Process Safety of AIChE.

CEN, EN 298, 1993
European Committee for Standardization, EN 298, *Automatic gas burner control systems for gas burners and gas burning appliances with or without fans*, 1993.

CENELEC, prEN50126, 1998
CENELEC, prEN50126, *Railway applications: The specification and demonstration of reliability, availability, maintainability and safety (RAMS)*, 1998.

CENELEC, ENV50129, 1998
CENELEC, ENV50129, *Railway applications: Safety related electronic systems for signalling*, 1998.

ESA, ECSS-P-001A, 1997
European Space Agency, ECSS-P-001A, *Glossary of terms*, Rev. 1, 1997.

Hölscher & Rader, 1986
Hölscher, H., R. Rader, *Microcomputers in safety technique: an aid to orientation for developer and manufacturer*, Verlag TÜV Bayern, 1986.

HSE, 1991
Health and Safety Executive, *Software for computers in safety related applications*, London, 1991.

IEC 50-191, 1990
International Electrotechnical Commission, IEC 50-191, *International electrotechnical vocabulary*; Chapter 191: 'Dependability and quality of service', 1990 (equal to BS 4778-3.2, 1991).

IEC 50-191, Am. 1, 1999
International Electrotechnical Commission, IEC 50-191, *International electrotechnical vocabulary*; Chapter 191: 'Dependability and quality of service', Amendment 1, 1999.

IEC 50-301, 1983
International Electrotechnical Commission, IEC 50-301, *International electrotechnical vocabulary*; Chapter 301: 'General terms on measurement in electricity', 1983.

IEC 50-351, 1998
International Electrotechnical Commission, IEC 50-351, *International electrotechnical vocabulary*; Chapter 351: 'Automatic control', 1998.

IEC 300-1, 1993
International Electrotechnical Commission, IEC 300-1, *Dependability management*; Part 1: 'Dependability programme management', 1993.

IEC 300-2, 1995
International Electrotechnical Commission, IEC 300-2, *Dependability management*; Part 2: 'Dependability programme elements and tasks', 1995.

IEC 300-3-1, 1991
International Electrotechnical Commission, IEC 300-3-1, *Dependability management*; Part 3: 'Application guide'; Section 1: 'Analysis techniques for reliability: Guide on methodology', 1991.

IEC 300-3-3, 1996
International Electrotechnical Commission, IEC 300-3-3, *Dependability management*; Part 3: 'Application guide'; Section 3: 'Life cycle costing', 1996.

IEC 300-3-4, 1996
International Electrotechnical Commission, IEC 300-3-4, *Dependability management*; Part 3: 'Application guide'; Section 4: 'Guide to the specification of dependability requirements', 1996.

IEC 300-3-9, 1995
International Electrotechnical Commission, IEC 300-3-9, *Dependability management*; Part 3: 'Application guide'; Section 9: 'Risk analysis of technological systems', 1995.

IEC 812, 1985
International Electrotechnical Commission, IEC 812, *Analysis techniques for system reliability: Procedure for failure mode and effect analysis (FMEA)*, 1985.

IEC 902, 1987
International Electrotechnical Commission, IEC 902, *Industrial process measurement and control: Terms and definitions*, 1987.

IEC 1123, 1991
International Electrotechnical Commission, IEC 1123, *Reliability testing: Compliance test plans for success ratio*, 1991.

IEC 1131-1, 1992
International Electrotechnical Commission, IEC 1131-1, *Programmable controllers*; Part 1, 'General information', 1992.

IEC 61508-4, 1998
International Electrotechnical Commission, IEC 61508, *Functional safety of electrical/ electronic/programmable electronic safety-related systems*; Part 4: 'Definitions and abbreviations', 1998.

IEEE Std-500, 1984
Institute of Electrical and Electronics Engineers, IEEE-Std-500, *Guide to the collection and presentation of electrical, electronic, sensing component, and mechanical equipment reliability data for nuclear-power generating stations*, 2nd edition, copyright © 1984. All rights reserved.

IEEE Std-610.12, 1991
Institute of Electrical and Electronics Engineers, IEEE-Std-610.12, *Glossary of software engineering terminology*, copyright © 1991. All rights reserved.

IEEE Std-729, 1983
Institute of Electrical and Electronics Engineers, IEEE-Std-729, *IEEE Standard glossary of software engineering terminology*, copyright © 1983. All rights reserved.

IEEE Std-730, 1989
Institute of Electrical and Electronics Engineers, IEEE-Std-730, *IEEE Standard for software quality assurance plans*, copyright © 1989. All rights reserved.

IEEE Std-829, 1983
Institute of Electrical and Electronics Engineers, IEEE-Std-829, *IEEE Standard for software test documentation*, copyright © 1983. All rights reserved.

IEEE Std-982.1, 1988
Institute of Electrical and Electronics Engineers, IEEE-Std-982.1, *IEEE Standard dictionary of measures to produce reliable software*, copyright © 1988. All rights reserved.

IEEE Std-990, 1987
Institute of Electrical and Electronics Engineers, IEEE-Std-990, *IEEE recommended practice for ADA as a program design language*, copyright © 1987. All rights reserved.

IEEE Std-1002, 1987
Institute of Electrical and Electronics Engineers, IEEE-Std-1002, *IEEE Standard taxonomy for software engineering standards*, copyright © 1987. All rights reserved.

IEEE Std-1008, 1987
Institute of Electrical and Electronics Engineers, IEEE-Std-1008, *IEEE Standard for software unit testing*, copyright © 1987. All rights reserved.

IEEE Std-1012, 1987
Institute of Electrical and Electronics Engineers, IEEE-Std-1012, *IEEE Standard for software verification and validation*, copyright © 1987. All rights reserved.

IEEE Std-1028, 1988
Institute of Electrical and Electronics Engineers, IEEE-Std-1028, *Software reviews and audits*, copyright © 1988. All rights reserved.

IEEE Std-1058.1, 1988
Institute of Electrical and Electronics Engineers, IEEE-Std-1058.1, *IEEE Standard for software management plans*, copyright © 1988. All rights reserved.

IEEE Std-1061, 1992
Institute of Electrical and Electronics Engineers, IEEE-Std-1061, *IEEE Standard for a software quality metrics methodology*, copyright © 1992. All rights reserved.

IEEE Std-1062, 1992
Institute of Electrical and Electronics Engineers, IEEE-Std-1062, *IEEE Recommended practice for software acquisition*, copyright © 1992. All rights reserved.

IEEE Std-1063, 1988
Institute of Electrical and Electronics Engineers, IEEE-Std-1063, *IEEE Standard for software user documentation*, copyright © 1988. All rights reserved.

IEEE Std-1074, 1991
Institute of Electrical and Electronics Engineers, IEEE-Std-1074, *IEEE Standard for developing software life cycle processes*, copyright © 1991. All rights reserved.

IEEE Std-1219, 1992
Institute of Electrical and Electronics Engineers, IEEE-Std-1219, *IEEE Standard for software maintenance*, copyright © 1992. All rights reserved.

IEEE Std-1228, 1994
Institute of Electrical and Electronics Engineers, IEEE-Std-1228, *IEEE Standard for software safety plans*, copyright © 1994. All rights reserved.

IEEE Std-1298, 1992
Institute of Electrical and Electronics Engineers, IEEE-Std-1298, *Software Quality Management Systems*; Part 1: 'Requirements', copyright © 1992. All rights reserved.

ISA, S84.01, 1996
Instrument Society of America, ANSI/ISA-S84.01-1996, *Application of safety instrumented systems for the process industries*, 1996.

ISA, S88.01, 1995
Instrument Society of America, ANSI/ISA-S88.01-1995, *Batch control*; Part 1: 'Models and terminology', 1995.

ISO/IEC, Guide 2, 1996
International Organization for Standardization, ISO/IEC Guide 2, *Standardization and related activities - General vocabulary*, 1996.

ISO 2382-1, 1993
International Organization for Standardization, ISO, ISO 2382-1, *Information technology - Vocabulary*; Part 1: 'Fundamental terms', 1993.

ISO 2382-14, 1997
International Organization for Standardization, ISO 2382-14, *Data processing - Vocabulary*; Section 14: 'Reliability, maintenance and availability', 1997.

ISO 2382-2, 1976
International Organization for Standardization, ISO 2382-2, *Data processing - Vocabulary*; Section 02: 'Arithmetic and logic operations', 1976.

ISO/IEC DIS 2382-7, 1996
International Organization for Standardization, International Electrotechnical Commission, ISO/IEC DIS 2382-7, *Information technology - Vocabulary*; Part 7: 'Computer programming', 1996.

ISO/IEC 2382-8, 1986
International Organization for Standardization, ISO 2382-8, *Information processing systems - Vocabulary*; Part 8: 'Control, integrity and security', 1986.

ISO 8402, 1994
International Organization for Standardization, ISO 8402, *Quality management and quality assurance - Vocabulary*, 1994.

ISO 9000-4, 1993
International Organization for Standardization, ISO 9000-4, *Quality management and quality assurance standards*; Part 4: 'Guide to dependability programme management', 1993 (equal to IEC 300, 1993).

Jones, 1992
Jones, *Nomenclature for hazard and risk assessment in the process industries*, Institution of Chemical Engineers, 1992.

Klinger et al., 1990
Klinger, D.J., Y. Nakada, M.A. Menendez, *AT&T reliability manual*, Van Nostrand Reinhold, New York, 1990.

Korteweg, 1994
Korteweg, L., *Classification and implementation of instrumented protective functions*, International symposium on PLCs in safety related applications, Cologne, 7-8 September 1994.

Laprie, 1992
J.C. Laprie (ed.), *Dependable computing and fault-tolerant systems*, Vol. 5, *Dependability: Basic concepts and terminology*, Springer-Verlag Wien, 1992.

Leveson, 1995
N.G. Leveson, *Safeware: System safety and computers*, Addison-Wesley Publishing Corporation, 1995.

Martz & Waller, 1982
Martz, H.F. & R.A. Waller, *Bayesian reliability analysis*, Wiley, 1982.

Musa et al., 1987
Musa, J., A. Iannino and K. Okumoto, *Software reliability; measurement, prediction, application*, McGraw-Hill, New York, 1987.

NATO, ARMP-7, 1996
North Atlantic Treaty Organisation (NATO), Military agency for standardisation (MAS), ARMP-7, *NATO R&M terminology applicable to ARMPs*, 1996.

O'Connor, 1981
O'Connor, P.D.T., *Practical reliability engineering*, John Wiley & Sons, Chichester, 1981.

OREDA, 1992
OREDA, *Offshore Reliability Data Handbook*, Published by OREDA participants, Pennwell Publishing Company, 1992.

RTCA, DO178B, 1992
RTCA, DO178B, *Software considerations in airborne systems and equipment certification*, Advisory Circular, 1992. Copyright © (1992) by RTCA and reproduced by permission of RTCA; reprinting information requires authorisation from RTCA.

SA, AS 3563, 1991
Standards Australia, Australian Standard 3563, *Software quality management system*; Part 1: 'Requirements', 1991 (adopted by IEEE as IEEE-Std-1298, 1992).

Smith, 1981
Smith, D.J., *Reliability and maintainability in perspective, technical, management and commercial aspects*, The Macmillan Press Ltd., 1981.

Smith & Wood, 1989
Smith, D.J. & K.B. Wood, *Engineering quality software*, Elsevier Applied Science, 1989.

SSCP 83, 1995
Sea Systems Controllerate Publication No. 83, SSCP 83, Issue 1.0, *MOD Project manager's guide to C and C++*, 1995.

SSCP 85, 1995
Sea Systems Controllerate Publication No. 85, SSCP 85, Issue 2, *Joint MOD/Industry computing policy for military operational systems*, 1995.

SSP 84, 1996
Sea Systems Publication No. 84, SSP 84, Issue 1.0, *MOD project manager's guide to ADA*, March, 1996.

UK DTI, ITSEC, 1991
UK Department of Trade and Industry, ITSEC, *Information Technology Security Evaluation Criteria*, June 1991.

UK MoD, Def Stan 00–41, 1993
UK MoD Directorate of Standardization, Defence Standard 00–41, *Reliability and Maintainability: MoD Guide to Practices and Procedures*, Issue 3 (supersedes all previously issued separate parts), June 1993.

UK MoD, Def Stan 00–49, 1996
UK MoD Directorate of Standardization, Defence Standard 05–49, *Reliability and maintainability: MOD guide to terminology definitions*, 1996.

UK MoD, Def Stan 00–55, 1997
UK MoD Directorate of Standardization, Interim Defence Standard 00–55, *Requirements for safety related software in defence equipment*, Issue 2, 1997.

UK MoD, Def Stan 00–56, 1996
UK MoD Directorate of Standardization, Interim Defence Standard 00–56, *Safety Management requirements for defence systems*, Issue 2, 1996.

UK MoD, Def Stan 00–58, 1996
UK MoD Directorate of Standardization, Interim Defence Standard 00–58, *HAZOP studies on systems containing programmable electronics*, 1996.

UK MoD, Def Stan 05–57, 1993
UK MoD Directorate of Standardization, Defence Standard 05–57, *Configuration management*, 1993.

UK MoD, Def Stan 05–95, 1995
UK MoD Directorate of Standardization, Defence Standard 05–95, *Quality system requirements for the design, development, supply and maintenance of software*, 1995.

USA DoD, MIL-Hdbk-272A, 1993
USA Department of Defense, MIL-Hdbk–272A, *Safety design and evaluation criteria for nuclear weapons systems*, 1993.

USA DoD, MIL-Hdbk-347, 1990
USA Department of Defense, MIL-Hdbk–347, *Mission-critical computer resources software support*, 1990.

USA DoD, MIL-Hdbk-505, 1998
USA Department of Defense, MIL-Hdbk–505, *Handbook for definitions of item levels, item exchangeability, models, and related terms*, 1998.

USA DoD, MIL-Hdbk-781A, 1996
USA Department of Defense, MIL-Hdbk–781A, *Handbook for reliability test methods, plans, and environments for engineering, development qualification, and production*, 1996.

USA DoD, MIL-Hdbk-1467, 1997
USA Department of Defense, MIL-Hdbk–1467, *Acquisition of software environments and support software*, 1997.

USA DoD, MIL-Std-109C,1994
USA Department of Defense, MIL-Std–109C, *Quality assurance terms and definitions*, 1994.

USA DoD, MIL-Std-280A, 1969
USA Department of Defense, MIL-Std-280A, *Definition of item levels, item exchangeability, models and related terms*, 1969 (redesignated as MIL-Hdbk-505, 1998).

USA DoD, MIL-Std-480B, 1988
USA Department of Defense, MIL-Std-480B, *Configuration control: engineering changes, deviations and waivers*, 1988 (cancelled by Notice 1, 1992).

USA DoD, MIL-Std-498, 1994
USA Department of Defense, MIL-Std-498, *Software development and documentation*, 1994 (cancelled by Notice 1, 1998).

USA DoD, MIL-Std-721C, 1981
USA Department of Defense, MIL-Std-721C, *Definitions of terms for reliability and maintainability*, 1981 (cancelled by Notice 2, 1995).

USA DoD, MIL-Std-785B, 1980
USA Department of Defense, MIL-Std-785B, *Reliability program for systems and equipment development and production*, 1980 (cancelled by Notice 3, 1998).

USA DoD, MIL-Std-882C, 1993
USA Department of Defense, MIL-Std-882C, *System safety program requirements*, 1993.

USA DoD, MIL-Std-973, 1993
USA Department of Defense, MIL-Std-973, *Configuration management*, 1992.

USA DoD, MIL-Std-2167A, 1988
USA Department of Defense, MIL-Std-2167A, *Defense system software development*, 1988 (cancelled by Notice 1, 1994).

VDE 0801, 1991
Deutsches Institut für Normung, DIN-V-VDE 0801, *Principles for computers in safety-related systems*, 1990.

MIX
Papier aus verantwortungsvollen Quellen
Paper from responsible sources
FSC® C105338

If you have any concerns about our products,
you can contact us on
ProductSafety@springernature.com

In case Publisher is established outside the EU,
the EU authorized representative is:
Springer Nature Customer Service Center GmbH
Europaplatz 3, 69115 Heidelberg, Germany

Printed by Libri Plureos GmbH
in Hamburg, Germany